PLACE IN RETURN BOX to remove this checkout from your record.
TO AVOID FINES return on or before date due.

DATE DUE	DATE DUE	DATE DUE

2/17 #20 White FORMS/DateDueForms_2017.indd - pg.1

Discover more in the series at
www.oxfordtextbooks.co.uk/obp

Published in partnership with the Royal Society of Biology

CONSERVATION

≋ OXFORD BIOLOGY PRIMERS

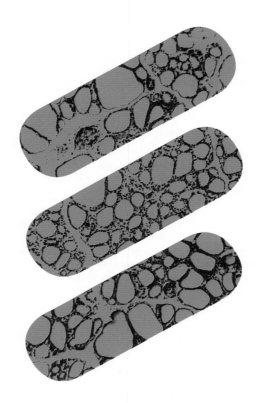

CONSERVATION

A people-centred approach

Francis Gilbert, University of Nottingham
Hilary Gilbert, University of Nottingham
Edited by Ann Fullick
Editorial board: Ian Harvey, Gill Hickman, Sue Howarth

OXFORD
UNIVERSITY PRESS

 Royal Society of
Biology

OXFORD
UNIVERSITY PRESS

Great Clarendon Street, Oxford, OX2 6DP,
United Kingdom

Oxford University Press is a department of the University of Oxford.
It furthers the University's objective of excellence in research, scholarship,
and education by publishing worldwide. Oxford is a registered trade mark of
Oxford University Press in the UK and in certain other countries

Published in the United States of America by Oxford University Press
198 Madison Avenue, New York, NY 10016, United States of America

British Library Cataloguing in Publication Data
Data available

Library of Congress Control Number: 2019935953

ISBN 978-0-19-882166-3

Printed in Great Britain by
Bell & Bain Ltd., Glasgow

PREFACE

Welcome to the Oxford Biology Primers

There has never been a more exciting time to be a biologist. Not only do we understand more about the biological world than ever before, but we're using that understanding in ever-more creative and valuable ways.

Our understanding of the way our genes work is being used to explore new ways to treat disease; our understanding of ecosystems is being used to explore more effective ways to protect the diversity of life on Earth; our understanding of plant science is being used to explore more sustainable ways to feed a growing human population.

The repeated use of the word 'explore' here is no accident. The study of biology is, at heart, an exploration. We have written the Oxford Biology Primers to encourage you to explore biology for yourself—to find out more about what scientists at the cutting edge of the subject are researching, and the biological problems they're trying to solve.

Throughout the series, we use a range of features to help you see topics from different perspectives.

Scientific approach panels help you understand a little more about 'how we know what we know'—that is, the research that has been carried out to reveal our current understanding of the science described in the text, and the methods and approaches scientists have used when carrying out that research;

Case studies explore how a particular concept is relevant to our everyday life, or provide an intimate picture of one aspect of the science described;

The bigger picture panels help you think about some of the issues and challenges associated with the topic under discussion—for example, ethical considerations, or wider impacts on society.

More than anything, however, we hope this series will reveal to you, its readers, that biology is awe-inspiring, both in its variety and its intricacy, and will drive you forward to explore the subject further for yourself.

ABOUT THE AUTHORS

Professor **Francis Gilbert** did his BA and PhD at St John's College Cambridge, then became a Junior Fellow at Gonville & Caius College and a Harkness Fellow in the USA before becoming a lecturer at Nottingham in 1984, where he has been ever since. He is an ecologist with two main interests: the conservation of South Sinai, where he has worked since 1986; and the biology of hoverflies. He has published almost 200 papers and 20 books, including books on the natural world for children of a variety of age bands, a popular account of the Bedouin gardens of South Sinai, a primer of hoverfly biology, and a guide to teaching quantitative biology. In 2004–7 he lived in Cairo and South Sinai in order to run a large project aimed at improving conservation across the Protected Areas of Egypt via research, monitoring, and public awareness. He is currently working on a monograph on the biology of hoverflies, a village history, and a guide to South Sinai.

Dr **Hilary Gilbert** did her BA in Modern & Mediaeval Languages at Girton College, Cambridge. She ran a Volunteer Bureau and Nottingham Community Health Council before becoming an NHS manager, leaving that to join the King's Fund in London. She then developed and ran Derbyshire Community Foundation for almost ten years before resigning to join Francis in Cairo, where she founded the Community Foundation for South Sinai while doing her PhD on the relationship between the Park and the South Sinai Bedouin living within its boundary. She now researches the health and wellbeing of the Bedouin, especially the women, while running the CFSS and its UK partner, the South Sinai Foundation. She has written about her work throughout her career, from research reports to having had a regular newspaper column in the Nottingham Evening Post. She has a number of published papers and is currently writing a book on her work in South Sinai.

ACKNOWLEDGEMENTS

We would like to thank Ann Fullick for inviting us to contribute to this series of books. She made many significant contributions by reading through several drafts of the entire book, and making important suggestions, additions, reductions, and changes, all of which have made the book so much better. The penultimate draft was also helped greatly by the comments of Sue Howarth. We are also extremely grateful to Markus Eichhorn, an ecologist who read the entire final draft and again made lots of important and very useful comments that significantly altered the final version.

We have shared much of our work in Sinai with Samy Zalat and a number of colleagues from Sinai, Egypt, and elsewhere, including the staff of the St Katherine Protectorate. In our research we are grateful for the support of the Universities of Nottingham and Manchester, the Leverhulme Trust, the Wellcome Trust, the Natural Environment Research Council, the British Council, and Suez Canal University.

In the ten years since we established the Community Foundation for South Sinai and the UK-registered South Sinai Foundation, our work would not have been possible without our Bedouin manager Mohammed Khedr al Jebaali, trustee Farag Mahmoud al Mas'oudi, women's advice workers Halima Khedr Ahmed and Jemi'a Fatih, and the support of all our Bedouin friends, especially Hussein Saaleh, Nasr Mansour Duguny, and Farhan Zidan.

CONTENTS

ABBREVIATIONS

°C	degrees Celsius
CE	common era
cm	centimetres
ha	hectares
ha^{-1}	per hectare
km	kilometres
km^2	square kilometres
m	metres

1 CONSERVATION, ECOLOGY, AND SCIENCE

Imagine that an alien spacecraft visited the Earth three million years ago, drawn to our extraordinarily blue planet (see Figure 1.1). This was the time when recognizably human apes first evolved, i.e. the genus *Homo*. What was the Earth like then?

The climate was a bit warmer than it is now, but had started cooling after the formation of the Greenland ice-cap. The flora and fauna were diverse and dense, with huge deciduous forests in temperate regions, and even larger coniferous forests in the colder north. South America had been isolated from the rest of the continents for millions of years but at this stage it was connected to North America by a strip of land rising up from the sea floor by the action of volcanoes. The event known as the Great American Interchange was in full swing, as many northern species invaded the south, and (fewer) vice versa. Large—really large—mammals were common everywhere, even in Australia where marsupials were the dominant mammalian group. The oceans were also warmer than now, and teemed with life, including many mammals (whales, sea cows, seals) and huge sharks.

Suppose these aliens revisited the Earth 2000 years ago. Would they have noticed much difference, apart from the slightly cooler climate? And suppose that they came again now: what would their main impressions be?

Two thousand years ago they would have been asking: where did all the large mammals go? Nearly all of them had become **extinct**—only a small fraction of the original diversity remained. The forests were all of a different composition, and much had been removed. And a new, influential

Figure 1.1 The Earth from space looks tranquil and stable. The ever-changing mass of different organisms just isn't visible until you spend time on the surface. It was this image from the Apollo 8 mission that brought home to people just how finite the Earth actually is.

NASA/Wikimedia Commons/Public Domain

mammal had appeared, one they had not seen before, a bipedal ape. This was us, *Homo sapiens*, and we had dispersed to all corners of the globe amazingly quickly.

And today, the aliens would be astonished. Humans have reached a total **biomass** greater than any single species in the history of the planet, and have modified the planet out of all recognition. Never mind the large mammals, where are *any* of the animals they recorded on previous visits? Many have disappeared altogether from most of the **habitats** they used to occupy. Since 1970, more than 80% of animals have been lost from areas where they once lived successfully, even if they have not become completely extinct, and densities of animal populations have been cut by more than 60%. Today it is **ecologists** who try to understand why these changes have happened, by exploring the **biodiversity** that is left, studying how it is maintained, and considering ways in which it can be conserved.

In this book we explore modern ideas in conservation by describing and developing three main take-home messages:

- many elements of the science of ecology are important because they underpin conservation practice today;
- despite the long history of ecological science, in many cases we have been looking at too small a scale (both spatial and also in time) to understand how biodiversity is maintained;
- the way people interact with their local biodiversity is a vital element of sustainable conservation.

Throughout the book we have illustrated the ideas using case studies mainly taken from our three decades of research work on the biodiversity and indigenous people of the St Katherine Protectorate in South Sinai (Egypt).

The problem

The big problem facing human beings, and all the other organisms on the planet, is that the ecological footprint of humans—in other words, the area of biologically productive land needed per person per year to sustain their lifestyles—exceeds the ability of the Earth to provide it. The total land required rose to more than one Earth in about 1986! The problem of providing enough resources for everyone is currently being solved by degrading the Earth's ecosystems to fuel our unsustainable habits. Of course, some countries use far more resources than others: North America, the EU countries, and Asia are all massively in deficit relative to what their land surface produces, while South America has the greatest surplus.

The pressure of the human ecological footprint leads to biodiversity loss and the need for conservation. What are the main causes of these losses? In 2000, Osvaldo Sala and his colleagues identified five sets of known drivers of biodiversity losses. In order of importance, these are:

- land-use changes causing habitat loss and fragmentation;
- human-induced climate change;
- the excess nitrogen deposited in soils from industrial and agricultural activities that encourages grass to dominate and crowd out wild flowers (concentrations have doubled);
- non-climate effects of increasing CO_2 concentrations in the atmosphere (such as the acidification of the seas—acid concentrations have increased by 35%);
- the impacts of human-aided invasions of alien species.

The impacts of changes in land use are likely to be 50% greater than those of climate change, and 100% greater than the other three drivers. We will look at the effects of land-use changes in Chapters 2 and 3, because it is relevant to the way conservation ideas have changed over the last 25 years. By the end of the book, you will understand what a huge problem change in land use really is, and just how difficult conserving biodiversity will be in the future.

The impact of climate change on animals and plants interacts with habitat loss and fragmentation. This is because the main effect of climate change is to shift the area of where any one species can live successfully. In a warming world, this habitable space moves either polewards across the landscape, to the North or South, or up in elevation, with species living higher up mountains than ever before. This happens because the area where the mean temperature is 15°C, for example, shifts in these directions under global warming. Survival then depends on whether a particular species can move, and if so, whether there is a suitable pathway for the movements to happen. Neither of these things can be assumed, and where habitats become too fragmented, a suitable pathway for organisms to move to other areas becomes less of a realistic possibility.

The underlying cause of all these drivers of biodiversity loss is us. The problem is the size of the human population, and the ecological pressure that this creates. In 2017 there were 7.5 billion of us; as countries get richer, so the average number of children per family drops as the investment of time and money in education and rearing becomes more important, eventually balancing the rate of mortality at an equilibrium population size. The best estimates suggest human numbers will reach an equilibrium at 10–12 billion in 2100, but they could go much higher. What will the Earth look like then?

Do we need all of the biodiversity we've got?

Does it really matter that so many species are being driven to extinction? After all, one mouse is very like another; one fly more or less similar to any other. Does it matter if one of them disappears forever (Figure 1.2)? To answer this, we need to understand what species *do* in ecosystems. What roles

Figure 1.2a and b Most of us value animals such as tigers, sloths, and elephants. Not so many people care about solitary bees or parasitic ticks—like this one you might find on your pet cat or dog. But does it matter if any—or all—of these animals die out?

(a)   (b)

© Anthony Short

do they play, and are these roles vital to the functioning of the ecosystem? These are big questions! Ecologists cannot pretend that they have all the answers yet, but it is certainly not for want of trying.

The species question

Before we get too involved in whether we need all species, we should discuss how many species there are and where they are found, and there is of course the thorny problem of what we mean by a 'species'. This is a traditional item for discussion by biologists, but non-biologists hardly ever realize that it is an issue. The simplicity of physics ('take an ideal gas . . .') is in stark contrast with the study of living organisms, which are by their very nature full of variables we can't control: individuals of one species differ in a multitude of ways (sex, age, size, experience, . . .), as do different species (size, structure, niche, evolutionary history, . . .).

So what is a 'species'? You probably have learnt that a species is a group of individuals that can interbreed to produce fertile offspring, but in the scientific literature there are at least twenty-four suggested definitions of what a species is. We can group them all using three kinds of criteria:

- Physical definitions use various types of measurable differences (morphology, DNA sequences, etc.) to separate different species.

- Biological definitions use reproductive barriers of one sort or another, or sometimes ecological differences (such as in the niche) to define a species; the 'interbreeding' definition is one of these.

- Phylogenetic definitions emphasize the process of evolving two species from one, or the continuity of populations through time.

For most complex multi-cellular organisms with two sexes, the usual biological species concept ('two species cannot interbreed') works most of the time, but there are lots of situations where it does not (asexual species, for instance). But you can see straight away why biology is such a complex science.

The role of a species in an ecosystem

The simplest way to describe what species do in ecosystems is the familiar 'pyramid of biomass' in Figure 1.3. This splits the biomass of an ecosystem or a community into a number of 'trophic levels'. There are usually only four of these: primary producers (plants), herbivores, primary carnivores, and top carnivores, sometimes with an extra layer to represent detritivores feeding on dead organisms. It is called a pyramid because this emphasizes the large size of the bottom level and the strongly reducing sizes of each succeeding level (although this is not always the case—it depends on the scale that you look at it). It is usually used to demonstrate the very approximate 90% loss of biomass going from one level up to the next, the so-called '10% efficiency rule'. Notice that the pyramid ignores what the species are, in favour of what they do—i.e. the overall feeding relationships within ecosystems. Despite their traditional appearance in school textbooks, ecological research does not often use the pyramid of biomass, reflecting

Figure 1.3 This pyramid of biomass shows the mass of different types of organisms in a marine habitat.

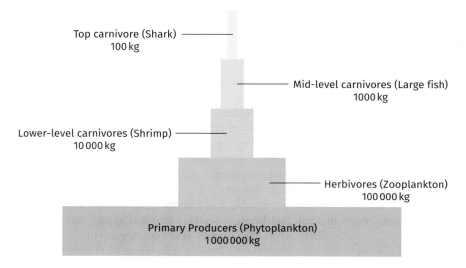

Top carnivore (Shark)
100 kg

Mid-level carnivores (Large fish)
1000 kg

Lower-level carnivores (Shrimp)
10 000 kg

Herbivores (Zooplankton)
100 000 kg

Primary Producers (Phytoplankton)
1 000 000 kg

the fact that too much detail is lost in its rather crude characterization of what is happening in an ecosystem.

An alternative model is the fully detailed food web, where every species in the ecosystem is identified, and its feeding relationships with other species recognized and quantified (for example, in terms of carbon or energy consumed per unit of time). These data can be represented in a food web, which can get very complicated, as in Figure 1.4. Imagine trying to obtain the basic information to draw such a diagram! Thus there are very few examples where literally every species is identified, simply because there are too many and identifying all of them is extremely hard (e.g. bacteria and protozoa). This is true even if the identification method is by sequencing all the DNA in a sample, because simple rules about identification using DNA sequence differences often do not work, again especially with organisms such as bacteria and protozoa.

Despite the difficulties, there are increasing numbers of impressive studies that identify the feeding linkages among sets of species, generally those within a defined habitat rather than entire ecosystems. Studying such food webs under different conditions can be illuminating. The rather beautiful example of Figure 1.5 shows the mutualistic interactions of ten of the most important forest trees/shrubs in Europe's last relict of ancient lowland forest, the great bison forest of Białowieża in eastern Poland. The webs show feeding relationships ('links') that are beneficial to the trees, i.e. insect pollinators and bird seed-dispersers. The comparison between the food web of the strict reserve of ancient forest with that of areas of the forest that have been logged in the past shows substantial losses of seed dispersal links, but an increase in pollinators.

There is a middle way between the simplistic pyramid of biomass and the extraordinary detail of species-based food webs. This keeps the structure of the web, but simplifies it by lumping species together into

Figure 1.4 This complex food web is based on a marine habitat with the cod as its focus.

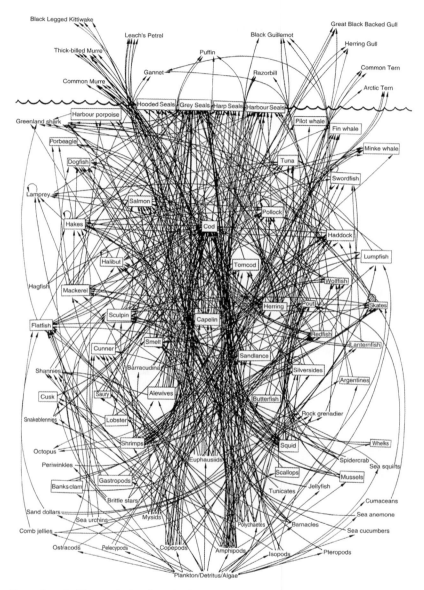

Reproduced with permission from Lavigne, D.M. 2003. Marine Mammals and Fisheries: The Role of Science in the Culling Debate, pp. 31–47, in *Marine Mammals: Fisheries, Tourism and Management Issues* (N. Gales, M. Hindell and R. Kirkwood eds.). Collingwood, VIC, Australia: CSIRO Publishing, 446 pp.

'functional groups', grouping together species that have similar ways of getting their resources. For example, it might split the herbivores into 'chewers', 'suckers', 'leaf miners' and 'stem borers', lumping together all the species that eat in those particular ways. Such trait-based ecological studies have revealed some important novel patterns in food webs.

Figure 1.5 Feeding relationships between ten common woody plants of the Bialowieza forest reserve of eastern Poland and their insect pollinators (blue) and seed dispersers (yellow, mostly birds). The width of the bars indicates relative abundance of each species. The upper diagram is for the core reserve of ancient untouched forest, while the lower diagram is for adjacent bits of the reserve which are managed forests and hence are logged. The species are *Rhamnus cathartica* (Purging Buckthorn), *Prunus padus* (Bird Cherry), *Euonymus europaeus* (Spindle), *Cornus sanguinea* (Dogwood), *Viburnum opulus* (Guelder Rose), *Sorbus aucuparia* (Rowan), *Frangula alnus* (Alder Buckthorn), *Rubus idaeus* (Raspberry), *Ribes nigrum* (Blackcurrant), and *Ribes spicatum* (Redcurrant).

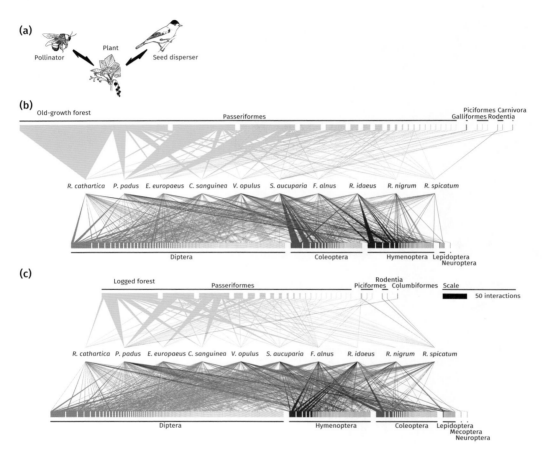

An example of this is a study of the food webs of soils, where there are many different functional groups, from predatory mites to bacteria-eating nematodes, shown in Figure 1.6a. Notice how the functional groups are a combination of taxonomic groups (eg cryptostigmatic mites [moss mites], nematodes) and feeding habits (e.g. predators, fungivores). This is not surprising since the most likely members of a functional group will be a set of related organisms that all get their food in a similar way.

How are these functional groups connected in the food web? The researchers measured rates of feeding for each predator–prey link in the food web, and also the impact of the predation on prey growth rates. Organizing the results by position in the web (see Figure 1.6b), an unexpected and non-intuitive pattern emerges, one of increasing importance as you go down in trophic level. The overall message here is that as you come down in trophic position in the food web, so the importance (in terms of influencing the growth rate) of the interaction to the prey increases. These results are novel outcomes of thinking about what organisms actually do in communities.

You can see that some of the less obvious interactions might be important to a food web. Experiments that remove some species or even an entire functional group are hard to do, especially on any but small scales, but nevertheless some such studies have been carried out. Removing single species only sometimes has major consequences. Thus advocates of 'rivet theory' might be right. Rivet theory compares the natural world to a machine held together by many rivets. As each rivet (species) is lost, nothing much seems to happen to the machine (the ecosystem functioning) until a critical point is reached. Then the loss of one more rivet results in catastrophic failure of the machine. Unfortunately for us, we cannot yet predict where the critical limit is.

Figure 1.6 1.6a shows a food web based on feeding relationships and diagram to show the feeding rate of different feeder-resource links (left) and importance of the link in terms of growth rate to the resource (middle) and the feeder (right). 1.6b shows how feeding relationships based on types of feeders rather than specific species give new insights into the relationships within an ecosystem. Food source (prey) on the left and predators on the right, listed from top of the food web down to the bottom. The bars indicate the rates upon which prey are fed (left), and the impact this has on their population growth rates (right). Notice how both measures increase as you go down the table (i.e. down in trophic position).

(a)

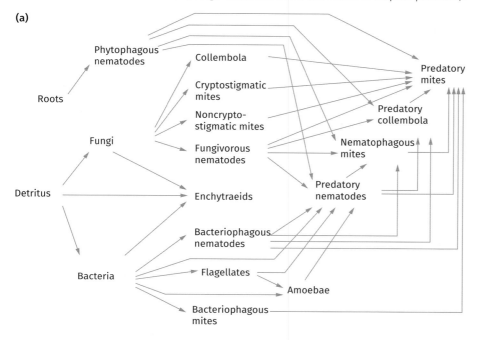

Adapted with permission from de Ruiter P.C., Neutel A.M., Moore J.C. *Energetics, patterns of interaction strengths, and stability in real ecosystems*. Science, Vol. 269, Issue 5228, pp. 1257-1260. Copyright © 1995, Copyright The American Association for the Advancement of Science. DOI: 10.1126/science.269.5228.1257

Figure 1.6 (Continued)

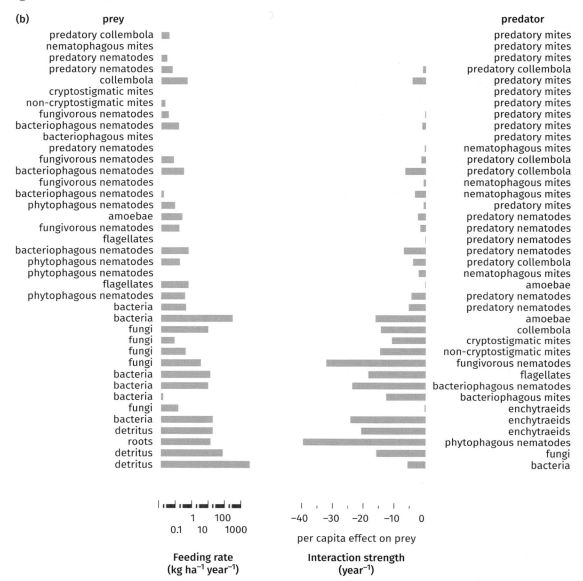

So, do we need all of the biodiversity we currently have in the world? For fairly short-term consequences to ecosystem functions, the answer appears to be 'no', or at least not always. It looks as if species which carry out the same function can substitute for one another if one becomes extinct, up to a point, although the system as a whole might become less able to withstand ecological shocks. The loss of genetic diversity might have longer-term consequences, but these are extremely difficult to test experimentally.

Hierarchy, scale, and explanation

Apart from dealing with living evolving organisms that are individually different, there are other important aspects of ecology (and hence conservation) as a science. Some of these are: (a) that it involves the hierarchical structure of nature; (b) that it involves huge changes of scale; and (c) that there are many different kinds of explanations for the same thing. These concepts are described briefly here to help you understand what comes later in this book.

Hierarchy and scale

In the natural world, molecules form cells, cells form organs, organs form individuals, individuals form populations, populations form species, species form communities, and communities form ecosystems. Now we also recognize two more components, metapopulation and metacommunity, inserted within this hierarchy. Having a hierarchical structure creates a contrast in the kinds of explanation that are acceptable in biology.

A long-standing problem in thinking about biology is the extent to which we should seek explanations of things that happen at one level in the hierarchy using properties *below* or *above* that level. These two opposing ways of doing science are 'reductionism' (from below) and 'holism' (from above). Look at Figure 1.7 for a summary of this idea.

Figure 1.7 To explain the physiology and behaviour of an organism such as this red-eyed tree frog from Costa Rica, should we look at what is happening at the level of the individual cells and molecules, or at the ecosystem of which it is a part? Reductionist and holistic scientists have very different opinions!

© Anthony Short

Reductionism is the standard approach of science in general, dominated as it is by ideas from the physical sciences. Under this view we expect the properties of a material to be explained in terms of the properties of the molecules it is made of. The properties of molecules can be understood from the nature of the atoms of which they are composed. So, for example, the way the nervous system works is then no more than the functioning of the cells, membranes, ions, and molecules that make it up.

But what if the way something works is a property of the context in which it is found? The same membrane molecule might have different functions in different cells. A hormone might have completely different effects in one organ as opposed to another. A species can invade one area but not another, depending on which species are already there. These holistic properties can be much harder to recognize and study. Some scientists do not like or do not recognize a holistic approach as a valid way of doing science, particularly those concentrating on working out the *mechanisms* of how things work (see later). However, in ecology it is much more of a possibility that the context is an integral part of how things work.

Holism interacts with the effects of scale. For example, if the 'community' or the 'ecosystem' has properties not explicable in terms of their constituent species and populations, then inevitably these properties occur at a much larger scale. In the 1970s and 1980s, some ecologists tried to deny such holistic properties in order to fit a reductionist approach. However, over the last 30 years ecologists have managed to show that the community context really matters for populations of animals and plants. To do this they have used many ingenious field-based experiments to make their point.

Scientific approach 1.1
Population growth and ecological arguments

It is easy to think scientific questions can be answered by simple experimentation. In many areas of biology—perhaps especially in ecology—this just isn't the case, and we have to use models. For example, after centuries of debate and experimentation, scientists still can't agree on what exactly controls the size of populations.

Scientists have been thinking about population growth ever since Thomas Malthus wrote 'An essay on the principle of population' in 1798. He claimed that the human population increases **geometrically** but agricultural production only increases **arithmetically**—and he predicted the result would be disastrous famine. Theoretically, if there are no resource limitations, populations do indeed increase geometrically because each individual female on average has more than one female offspring, resulting in an exponential increase in population size. This overproduction of offspring is one of the principles of Darwin's theory of natural selection.

What prevents populations from growing indefinitely? As Malthus recognized, food resources are a major requirement, and these are not infinite. But is the availability of food and other resources *always* the main reason for population limitation? There was a huge argument among ecologists about this in the 1920s, 1940s, and again in the 1970s—and it is still not resolved. The dispute was about whether a **density-dependent** effect is always needed in order to understand why populations are limited—for instance, whether population size is always the result of the balance of the numbers of animals and the food resources available. Some ecologists, often entomologists studying insects, claimed that bad weather alone could limit the size of a population—a cold, wet summer can greatly reduce the population sizes of organisms from wasps to butterflies and moths. Other ecologists (mostly those studying vertebrates) insisted that only density-dependent effects of lack of resources and/or effects of **natural enemies** can **regulate** population numbers.

Logically, in order to regulate a population, there might be a higher chance of dying (a negative effect) when the population is at a high density because resources are spread more thinly. This negative effect will get stronger the higher the density gets. In the same way, there might be more births per female (a positive) when population densities are low and there are lots of resources to go around. This positive effect will get stronger the lower the density gets. This is a nice story, but the evidence doesn't always support this model.

There are a lot of data from the field, and thousands of theoretical models. One of the longest datasets is the 200 years or more of records of the furs of snowshoe hare, lynx, and other animals trapped in Canada and traded to the Hudson Bay Company. This famous example shows strong population cycles (see Figure A). However, virtually all the models created to understand population dynamics were of single *closed* populations. The changes in the populations therefore depended only on births and deaths, and not at all on movement of individuals. Virtually all of the field data were limited to extremely small areas, simply because counting organisms in small areas is so much easier than counting them across large areas. As a result, neither the models nor most of the field data were at a large-enough scale to contain the processes that regulate populations, which happen at much larger scales than a single rose bush, or a single field or wood. No wonder there were arguments!

It was not until the 1970s to 1990s that the blinkers were lifted from our eyes, when the role of movement in population sizes became appreciated. The vital importance of movement became obvious from studies of islands, where chance and bad luck could cause a population to suffer an extinction, but this could be followed later by a recolonization from the mainland. These studies were mostly focused on the number of different species living on islands (see Chapter 2), but the findings were extended later to give a better understanding of changes in the size of the populations of individual species as well.

This renewed attention on movement and population extinctions allowed both theoretical and field biologists to expand their scope to study networks of patchy populations connected by movement. So when the spatial position of the Canadian fur returns was included in the dataset, an extraordinary pattern was revealed (see Figure B). Snowshoe hare numbers fluctuate in their familiar cycle in any one place—but they also fluctuate across the Canadian landscape just like the waves on a pond caused by dropping a stone. The waves of high and low numbers move outwards from a central point in western Canada. This kind of large-scale pattern can only be generated by movement of organisms into or out of an area.

SA 1.1 Figure A Numbers of snowshoe hare and lynx from the fur returns of the Hudson Bay Company.

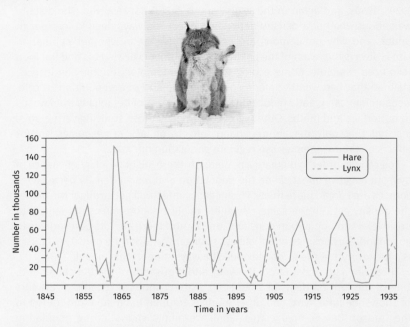

SA 1.1 Figure B The country-wide ripples of lynx populations across Canada. This is schematic—the real ripples are much more complex as they flow around objects in the landscape (such as lakes) (constructed from data in Smith & Davis 1981 J Biogeography 8(1): 27-35).

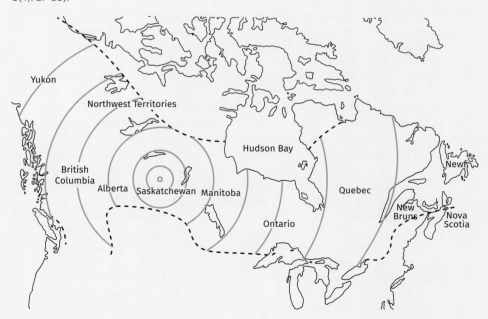

Adapted with permission from data from Smith, C. et al. A Spatial Analysis of Wildlife's Ten-Year Cycle. *Journal of Biogeography.* Copyright © 1981, John Wiley and Sons. [DOI: 10.2307/2844590]

Unlike many lab-based investigations, notice just how long it takes for the science of ecology to realize its mistakes and change. This is because the dynamics of populations happen over years and decades rather than minutes or days. Instead of being able to test a theory within a single day of experiments in the lab, it can take decades before an ecological answer becomes clear.

❓ Pause for thought

We have looked at a situation where ecological investigation of a question starts with the simplest data and models, and it takes decades for a definitive answer to emerge. Are there any other fields of biology like this?

Scientific explanations

We end with how we explain things in science. Even the Greeks knew that there were several completely different ways of explaining how something works, two of which they called the 'efficient' and the 'final' causes. We know these today as the 'mechanical' and the 'functional' ways of explanation. In biology, however, there are more than two ways of explanation.

For the question 'Why are swans white?', there are four possibilities, known as 'Tinbergen's Four Why's', after their first formulation by Niko Tinbergen, a Dutch biologist studying behaviour.

- The **mechanistic explanation** talks about the pigment and structural chemistry of the feathers that lead to the white colour.
- The **ontogenetic explanation** talks about the development of the feather colours from the egg to the adult, ending with the white colour.
- The **functional explanation** talks about the evolutionary costs and benefits of white feathers, testing if we know why white is the best colour for a swan.
- The **phylogenetic explanation** involves the evolutionary history of swans and where in that history the feathers evolved to become white.

These are all causal explanations of why swans are white, and all are equally valid—but it is obvious that they are very different. Thus it is important to recognize what kind of explanation is being put forward. So, for example, when a reductionist says that all biology is 'really chemistry', or all chemistry is 'really physics', you can see that they are offering only mechanistic explanations. It is certainly true that biology has unusually complex causality!

Ecology is therefore a field of science where hierarchy and scale are really important, and also where one must be clear about the kind of explanation being put forward.

❓ Pause for thought

Can you think of a similar 'why' question: what four possible explanations are there for your question?

Ecology and evolution

Evolution by natural selection was a truly revolutionary breakthrough in our understanding of the principles of biology, so much so that we are still realizing and working out its more subtle consequences. It is one of the simplest of scientific theories, but also one of the most misunderstood. It is deceptively simple—many think they understand it, but in reality its principles and implications are not straightforward at all. Charles Darwin is justly celebrated as its discoverer—it really is one of the most important ideas ever discovered. At its fundamental core there are just four principles, and a conclusion:

(i) *overproduction*: more offspring are produced than the environment can support;

(ii) this means there is *competition* for the resources needed to survive and reproduce;

(iii) *variation*: the traits of individuals vary in ways that influence the outcome of this competition; and

(iv) at least some of these differences are *heritable*, i.e. they have a genetic basis.

If all of these things are true, then there will be selection pressure leading to adaptation and evolution by natural selection. Evolutionary change can be defined as changes in the relative frequencies of alleles in a population, and if this results in an improvement to the average fitness, then adaptation has occurred. Fitness is ecological 'success', usually measured in an individual as the number of their offspring that survive to reproduce in the next generation.

Remember that other processes can also produce evolutionary change in the sense of changing allele frequencies between generations, but they cannot produce adaptation. These other processes include mutation, the immigration of individuals into the population with new alleles, and genetic drift. They usually erode the adaptation of a population to local conditions, but can sometimes produce new, better combinations and variants that can then be selected for and hence spread through a population.

Different fields have been revolutionized by the theory of evolution by natural selection at different times as its full implications have sunk in. Population genetics was first in the early twentieth century, and then taxonomy and systematics in the 1950s. Animal behaviour was not until the 1960s, ecology in the 1970s, and psychology and medicine started to be revolutionized in the 1990s and the process is still unfinished. Finally, the Darwinian revolution in the social sciences did not begin until the twenty-first century.

Every now and again, someone claims to have discovered that something extra is required to the theory of evolution by natural selection, but such 'new' ideas have always faded away after having been shown to be just natural selection from a slightly different angle. One of the few real additions was made by Darwin himself in the idea of sexual selection—males compete to be chosen by females to fertilize their eggs. This can generate counter-intuitive selection, resulting in extravagant male structures such as the peacock's tail. Another was also thought of first by Darwin—the idea of inclusive fitness. The success of an individual includes the offspring of others (normally relatives) that share the same genes.

Ironically, Darwin's title, *'On the Origin of Species by Means of Natural Selection'*, was the very bit of the process he never did understand—i.e. speciation. He showed that natural selection can cause changes, but he then assumed that natural selection + millions of years = biodiversity. At the time no-one knew anything about the role of genes in inheritance, nor anything at all about how speciation happened. We now know a huge amount more, but there is of course plenty more yet to discover.

The speed at which evolution happens is traditionally thought of as slow, of the order of millions of years. We do know that the genetic differences between the most closely related species, 'sister-species', are often not large, but can be concentrated into particular bits of some chromosomes. These differences normally develop between populations living in different areas, ones that are not in contact with one another. It is only if and when these populations recontact one another that it becomes clear whether they are now different species or not. If they can interbreed, with fertile hybrids, they are (usually) still the same species.

Sometimes two populations are sufficiently different for the hybrids to be relatively non-viable or infertile, then there is rapid selection not to interbreed: this is called 'reinforcement'. With reinforcement, two species can result after as little as a few thousand years. Without any contact, and hence no reinforcement selection, two gradually diverging populations can become different species in about 2.7 million years. However, these are estimates derived from data on fruit flies, and speciation rates vary greatly between different types of organisms. Species which breed rapidly and have very short life spans can evolve much faster than large organisms that sometimes only have offspring every few years.

What about evolutionary change that does not involve speciation? How fast is that? We now realize that such changes can be rather fast. As an example, we can look at stickleback evolution. The Three-spined Stickleback (*Gasterosteus aculeatus*) is one of the main model organisms used to study evolutionary changes because, although largely marine in the northern hemisphere, it has repeatedly invaded freshwater habitats. Most populations live in the sea, but breed in fresh or brackish water; to do this they need to be very tolerant of changes in salinity. Thus if fish in a coastal pond are cut off from the sea by an earthquake, for example, their tolerance to low salinity makes survival possible. All over the world there are freshwater populations derived from such invasions from the sea. Marine and freshwater forms are often starkly different in size, shape, and structure, especially in the bony armour plates with which they are clad (see Figure 1.8). In marine forms the bony plates are strongly expressed, whilst they are lacking, or small and weak in freshwater forms.

On 27 March 1964, the largest earthquake ever recorded in North America struck the south coast of Alaska. It measured 9.2 on the moment-magnitude scale (for comparison, the 2011 Fukushima earthquake with its huge tsunami measured 6.9 on this scale). The Alaskan earthquake uplifted islands by several metres within minutes, creating terraces with new ponds from previously submarine platforms. Many of these ponds contained stickleback, and these populations now look like stickleback from ponds established thousands of years ago. The morphological and accompanying genetic differences have evolved in only 50 years, rather than gradually over thousands of years. Whether this involves speciation, or not, is debatable.

Figure 1.8 Differences in appearance between marine (bottom) and freshwater (top) forms of the Three-spined Stickleback, *Gasterosteus aculeatus*, photographed together in an aquarium. Apart from the obvious size difference, the bony plates of the marine form are lacking in the freshwater form.

Bony plates

© Andrew MacColl

Ecologists used to think that ecological timescales of days, weeks, months, and years were very different from the timescales of hundreds or thousands of years typically thought to be required for evolutionary changes. But the example of the stickleback shows that this is not the case. Evolutionary change can happen in really short time periods that overlap with ecological timescales. Substantial evolutionary change can even happen under human-induced selection pressures, such as climate change or agricultural land use.

In a recent example scientists looked for genetic differences among eight common grasses growing in a range of land-use types (mowing, grazing, and soil fertilization regimes) at different intensities in 150 sites in Germany. The regimes had been similar for at least ten years in each site—a mere instant in time if we are thinking of evolutionary timescales. Seeds of individual plants from these sites were grown in a common environment to reveal if there were genetic differences caused by the different regimes. A set of traits were measured, such as when they flowered, how large they were, the number of flowers, etc. The researchers were surprised to see strong genetically based differences among the land-use types, the result of evolved changes. Mowing selected for early flowering, soil fertilization for different flower shapes and sizes according to species, while grazing selected for decreased biomass and late flowering.

Thirty years ago, no ecologist would have believed that such evolutionary effects were possible over such a short time frame. This shows us that even short-term management of habitats can cause evolutionary changes in animals and plants, and we need to bear this in mind when we try to conserve species or habitats.

Case study 1.1

The Large Blue butterfly *Maculinea arion* (Lepidoptera: Lycaenidae)

Butterflies and moths (the Lepidoptera) form one of the four main orders of insects: the others are the flies (Diptera), the beetles (Coleoptera), and the ants, bees, and wasps (Hymenoptera). All of these groups have four stages to their life cycle: the egg, larva (which looks nothing like the adult), pupa (when the larval body is totally reorganized), and the adult.

The extinction rates of butterflies are higher than those of vertebrates or plants, and extinctions have occurred in nature reserves even where their larval and adult food plants are abundant. Attempts at conserving such butterflies often failed because the cause of the decline was not understood. The UK's population of Large Blues declined from 91 colonies in 1800, to 25 colonies with tens of thousands of adults in 1950, to two colonies with about 325 butterflies in 1972, before national extinction in 1979. Why did this happen, despite every attempt to save it?

The Large Blue is an extreme specialist whose larvae switch from a brief three-week period feeding on a plant (thyme) to living as a social parasite inside the nests of red ants of the genus *Myrmica*. The butterfly occurs only in herb-rich south-facing grasslands where there are lots of thyme and *Myrmica* ants. The 4th-stage larva abandons feeding on the flower buds of the thyme plant, and wanders until it is adopted into the underground nest by the first *Myrmica* worker to encounter it. This happens because the chemistry and behaviour of the larva mimic those of an ant grub, and the worker picks it up and 'restores' it to the brood chamber. The larva then eats ant brood (eggs, larvae, and pupae) for ten months until it pupates and forms an adult butterfly. You can see the life cycle in Figure A.

It took six years of fieldwork (1972–8) to discover the answer: can you imagine the effort involved? Analyses and modelling of the data identified the mortality of the 4th-stage larva as the key factor determining overall population changes in the Large Blue. There were four causes of this mortality:

(i) survival depended on being encountered by the right species of ant—there were five species of *Myrmica* foraging under thyme plants, but the chance of survival of a Large Blue larva was more than five times higher if a worker of *Myrmica sabuleti* adopted it; the larvae all die if there are no *Myrmica* ants available;

(ii) the presence or absence of queen ants—these are needed to maintain the supply of ant brood;

(iii) the density of Large Blue larvae in the nest—a larva can starve to death if there are too many others in the nest eating the ant brood;

(iv) very dry weather conditions—mortality increases in a drought year.

Adult female butterflies suffer high predation from birds and invertebrates, and on average they live only for a few days (mean 3.9 days). In a

CS1.1 Figure A The life cycle of the Large Blue butterfly is very specialized and until we understood it fully, all efforts to conserve this beautiful creature failed.

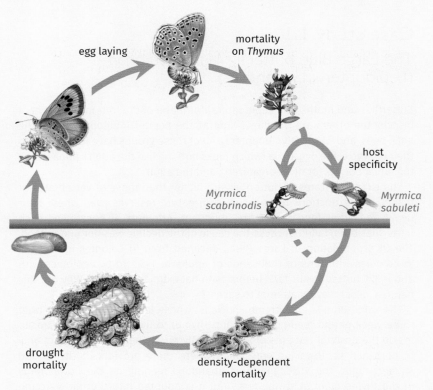

Reproduced from J. A. Thomas, D. J. Simcox, R. T. Clarke. Successful Conservation of a Threatened *Maculinea* Butterfly. *Science* 325: 80. Copyright © 2009, American Association for the Advancement of Science.

typical year each female lays an average of only 25.5 eggs, even fewer during a prolonged wet summer (11.5) or a summer of extreme drought (8.0). Captive females can live much longer because they have no predators— they can lay more than 200 eggs.

An equation (or 'model') containing all these factors could predict 76% of the variation in population numbers in the last few places where the butterfly occurred before its extinction. This is an extremely good model, and therefore it was used to estimate what the best conservation strategy for reintroduction would be. Clearly the numbers of *Myrmica sabuleti* were the critical factor, and the model suggested that if more than two-thirds of larvae were adopted by this species, then butterfly numbers would increase.

Surveys of every known former site finally enabled the cause of the decline to be identified. People had suggested a number of factors (over-collecting of adult butterflies, grassland patch size, or patch isolation) that might be the cause of the problem, but the research showed none of these had anything to do with the extinctions of the butterfly. Instead, the availability of *Myrmica sabuleti* was the driver. The presence of this ant depended on the quality of the grassland habitat, and was closely connected with the height of the grass turf (see Figure B); areas of grass less than 1.4 cm tall were needed to encourage *Myrmica sabuleti* to be the dominant ant (i.e. to be more than two-thirds of workers foraging under thyme). This ant needs hot conditions, and if the grass grows to be more than 2 cm high, the temperature in the ant brood chamber falls by 2–3°C. This small change makes the difference in competition between *Myrmica sabuleti* and its less heat-loving relative *Myrmica scabrinodis*, and the latter becomes the dominant ant—which does not help the Large Blue larvae.

As soon as this key relationship had been identified, it was obvious what had happened. The hills had long been grazed by sheep, which kept the grass very short, allowing *Myrmica sabuleti* to be the dominant ant and the Large Blues to thrive. But sheep farming became uneconomic, and sheep numbers fell, leaving only the rabbits to keep the hillside grasslands short. Then in the 1950s, the rabbit disease myxomatosis arrived in the UK, deliberately introduced to get rid of the excessive numbers of rabbits. The grazing pressure disappeared along with the rabbits—all but one site became too overgrown for the red ants, *Myrmica sabuleti*, and hence the Large Blues, to survive.

Unaware of this, early conservationists, including the Royal Entomological Society, did exactly the wrong thing. For example, they erected fences to deter collectors and hence prevented a local famer from 'swaling' (burning gorse) and grazing the site. As a result, the grass grew long and the Large Blues there rapidly died out. By the time the cause of the decline had been identified, it was too late and the last UK population was extinct. The only thing to do was to reintroduce them from suitably similar populations from Sweden.

Sites were carefully prepared by reintroducing grazing and making sure there was a network of suitable sites within dispersal range (c. 250 m). The grazing had the desired effect in that *Myrmica sabuleti* once again became the dominant ant. Three of four initial reintroductions were successful, then a further four, and by 2008 the butterfly had naturally colonized 25 more sites. It spread step by step, occupying neighbouring habitat patches first. It took 14 years to reach the furthest patch 4.4 km from one of the reintroductions. The rate of spread increased with time, probably as a result of selection for dispersive adults causing them to have a slightly larger thorax with more powerful flight muscles.

In introduction sites in Somerset, where local climates were about 1.5°C cooler than the source sites in Sweden, the synchrony between adult butterflies and flowering thyme was imperfect but adequate. Slightly further north in the Cotswolds, where temperatures were half a degree cooler than in Somerset, it was not: the introductions failed because adult butterflies emerged too late to synchronize with the production of flower buds by thyme plants, leaving adults to lay eggs on later-flowering plants growing in the coolest spots—the only places where *Myrmica sabuleti* was absent. It was an **isotherm** too far. Later on, after 18 generations of selection to UK conditions in Somerset, the synchrony was excellent and a Cotswold reintroduction from Somerset in 2010 was a great success.

CS1.1 Figure B Relationship between turf height and the dominance of the *Myrmica* ant community by *Myrmica sabuleti*, the ant crucial to the survival prospects of Large Blue caterpillars, in sites in the UK. Blue symbols are sites with Large Blues, red symbols are sites where it became extinct, and black symbols are other sites. The solid red line describes the relationship established in the left-hand panel. The transverse dotted line is the predicted proportion of *Myrmica sabuleti* that allows Large Blue populations to sustain themselves.

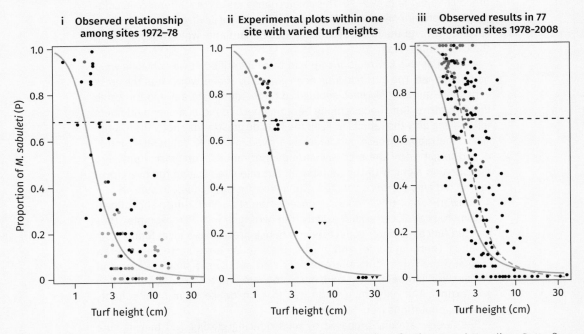

Reproduced from J. A. Thomas, D. J. Simcox, R. T. Clarke. Successful Conservation of a Threatened *Maculinea* Butterfly. *Science* 325: 80. Copyright © 2009, American Association for the Advancement of Science.

The largest populations of Large Blues now contain up to 5000 adults per hectare, ten times greater than any previously recorded population anywhere in the world. The model again works extremely well in predicting the effect of turf height on population sustainability (Figure Biii), although interestingly the minimum threshold turf height for the Large Blue is now a bit longer at 2.1 cm—largely explained by climate warming reducing the shading effect of the vegetation. The model also works superbly well in predicting the population dynamics of the butterfly—Figure C shows the trajectories for two sites, and the fit is remarkably close.

The successful conservation of the Large Blue is an outstanding example of just how detailed the knowledge of the precise requirements of this species needed to be. Luckily not every species is quite this hard. Notice how long it took to work out what these were. The discovery of the relationship with ants was made in 1910–1920, then six years of skill, effort, and insight in the 1970s were followed by three decades of careful work to build up the ant populations, reintroduce the butterfly, and follow up on the successes and failures. A set of connected sites was vital for conservation success, rather than just one site. There was also

CS1.1 Figure C Predicted (dashed) and actual (solid) population numbers for two populations (orange and red) of Large Blues after land management had restored the populations of the crucial ant *Myrmica sabuleti*.

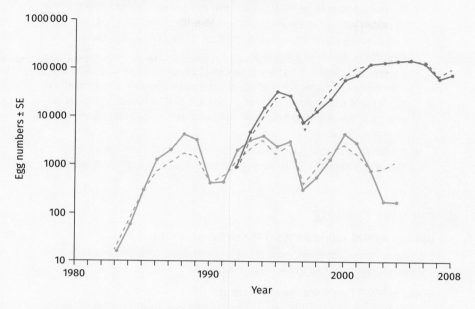

Reproduced from J. A. Thomas, D. J. Simcox, R. T. Clarke. Successful Conservation of a Threatened *Maculinea* Butterfly. *Science* 325: 80. Copyright © 2009, American Association for the Advancement of Science.

an important role for evolutionary change in both dispersal ability of the butterflies and synchrony of the life cycle of the butterflies with the thyme which acts as host-plant for the eggs.

Notice also how the roles of each species all need to be aligned in order for the Large Blues to succeed. In terms of the grazers, any species will do as long as the resulting sward is the right height: sheep, goats, cattle, or rabbits—any will do the job. However, do we need all the species of *Myrmica* ants, when all seem to be doing more or less the same thing in the habitat? Well, we certainly need *Myrmica sabuleti* if we want Large Blue butterflies!

❓ Pause for thought

This chapter discusses the role of scale in ecology: what role do scale effects play in the case study of the Large Blue? What is their importance in the long term?

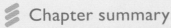 Chapter summary

- The environmental crisis generated ultimately by the demands of the increasing human population will drive many species to extinction. Whether most of these extinctions matter depends on what the species actually do in ecosystems. It's hard to be sure which species are important, and which are not.
- Ecological science is distinguished by the especially strong roles of the biological hierarchy and of spatial and temporal scale, as well as the kind of explanation that we seek. Over the 150 years of ecological research, much of the fieldwork has been at too small a scale, either in space or time.
- Evolutionary change happens much more quickly than we once thought, interacting with ecological time scales in important ways.

Further reading

Dawkins R (1976, reprinted 1989) *The selfish gene*. **OUP.**
A hugely important book, even though out of date in places, because it shows the right way to think about evolution.

Gould SJ (1977) *Ever since Darwin*. **Norton.**
A wonderful set of articles about case studies in evolution.

These two websites give further details about the case study:
UK Butterflies: Large Blue *www.ukbutterflies.co.uk/species.php?species=arion*
CEH: the Large Blue butterfly *www.ceh.ac.uk/large-blue-butterfly*

Discussion questions

1   Which of Tinbergen's Four Why's do you think is the most important to conservation, and why?
2   What are the most important ecological principles illustrated by the successful reintroduction of the Large Blue to the UK?

2 POPULATIONS, PATCHINESS, AND MOVEMENT

We now turn to think about populations in more detail. We describe the basic elements that help us to think about the way population numbers change, then look at how ecologists have described the causes of changes, and how population numbers are regulated in relation to population density.

Figure 2.1 A male Red Deer in the Peak District in winter.

© Andrew MacColl

We then consider the role of patchy habitats. Our ideas about this were transformed by the study of islands, and by a very influential theory about the numbers of species on them. This theory switched our focus from density-dependence to more of a focus on movement.

Changes in population size

In ecology, we often consider the number of individuals in a population. This sounds simple, but we immediately run up against a major issue. What is an individual? This seems easy if you are thinking about most animals, but what about plants such as grasses? Most new grass is not produced from seed, but from underground runners from one plant that produce new shoots every so often as they grow. Are these the same individual, or different ones? And what is a 'population'? What determines its boundaries? In ecological fieldwork, counts can be made of individuals within a field, for example, but is this a 'population'? The practical problem of defining boundaries is always difficult, and is to some extent always arbitrary.

When we say the population size of an organism is 343, it is easy to assume that all the members of a population are identical, all milling about in the habitat. But individuals are not at all alike.

The members of a population differ in a variety of ways. For instance, most animals come in two different sexes. Plants are usually hermaphrodites because they have both male (stamen) and female (pistil) parts, but some plants also come in two different sexes, often with the female flowers on one plant and the male ones on another. Individuals also differ in age or stage. Many animals pass through a set of stages in their life cycle (e.g. egg, larva, pupa, adult). At any particular time, a population is a mixture of these different sexes, ages or stages (Figure 2.2).

Individuals in and individuals out

When thinking about changes in population size through time, we are interested in the rate of reproduction because that is one of the processes that brings new individuals into the population. A population of males, or a population of juveniles, would be very different in its rate of reproduction from a population of mature mated females. Thus the sex ratio and age or stage structure of a population really matters.

In population dynamics, we only consider individuals that are capable of producing offspring, because only they matter in creating the next generation. So, for example, the size of an animal population usually refers just to the number of females. This is a reasonable thing to do because of a very basic asymmetry: in nature, virtually all females that survive to reproductive maturity get mated, but not all males get matings. Male–male competition generates sexual selection. In the extreme, for example in Elephant Seals (*Mirounga* spp) or Red Deer (*Cervus elaphus*; Figure 2.1), just a few

Figure 2.2 If you are not a keen naturalist, you might never realise that these two holly trees are members of the same population – or that the caterpillar and the butterfly are also members of the same population, just at different life stages.

(a) FloralImages/Alamy Stock Photo, (b) MichaelGrantPlants/Alamy Stock Photo, (c) jps/Shutterstock.com

males gain nearly all the matings. So it is reasonable to count only mature females, because as long as there are *some* males, then the females will be able to reproduce. It is extraordinarily rare for females to fail to reproduce because they are not mated.

The basic ecological fact of life is that a change in numbers in a population between two time periods occurs via four processes:

- births and immigrations add individuals to the population
- deaths and emigrations remove individuals from the population

$$N_{now} = N_{before} + births + immigrations - deaths - emigrations$$

These are the demographic processes, and one of the main aims of the science of ecology is to describe, understand, and predict these processes.

The way births are typically related to age (Figure 2.3a) means that there are no births at all until a female is mature, then a rapid rise to a peak, followed by a gradual decline until the cessation of reproduction (called the 'menopause' in humans). Nearly all female organisms die before reproduction ceases. The shape of the relationship between survival and age (Figure 2.3b) is also very important because it determines what the strategy of a female should be to get the maximum number of her offspring into the next generation. The three main types are shown in Figure 2.3b. Type I is typical of humans because it denotes an organism with a low mortality rate and hence a high chance of surviving to reproduce and into old age—it is only in old age that the chances of survival drops. In this case, females can be confident of being able to reproduce as much as they like, hence delays in reproducing until relatively late in life are possible options. Type II is where the mortality rate is a bit higher, so the chance of surviving declines in a straight line: many mammals and birds are like this. Type III is where juvenile mortality is very high, so the chance of surviving is low. Most insects and many fish are like this. Here the best strategy is to mature as fast as possible, then reproduce as fast as possible, since the chances of surviving to the next time period are slim. Notice that the differences between these types are related to body size: the larger an organism is, the more like a Type I they are.

Now take a look at Figure 2.4a–d. This shows some examples of counts of animals through time from several different studies. Notice how completely different they are! The numbers of sheep in Australia shows a steady increase until about 1850, followed by relatively small fluctuations around roughly the same mean value of about 1.6 million sheep. In the Canadian fur data from the Hudson Bay Company, the numbers of lynx show very regular cycles of 10.8 years, but there are extremely large changes in numbers from peak to trough, and the magnitude of this change is very variable. Lemming numbers also fluctuate, but much more irregularly, while the numbers of thrips (insect pests) on roses fluctuate apparently at random, although the variation among the summer peaks and the variation among the winter troughs are both small relative to the within-year variation. The populations do not die out, and neither do they increase for ever.

Figure 2.3a and b Different organisms have evolved different strategies to maximize their reproductive success and these are affected by age: (a) shape of the curve describing births (the number of offspring) produced at different ages for a typical organism; (b) shape of the curve describing the chances of surviving (the inverse of mortality) to different ages for three different kinds of organism.

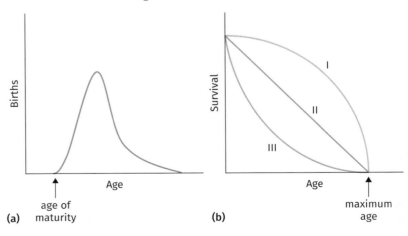

Figure 2.4a–d Different kinds of population dynamics (from Begon et al).

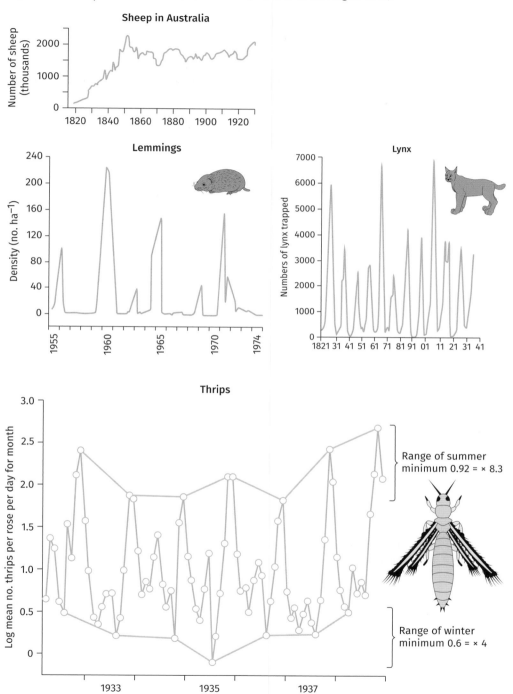

So why *do* population numbers change?

Ideally we would like a single theory to explain all the different patterns of population numbers seen in Figure 2.4, and in particular to explain how and why numbers do not increase exponentially and do not reach zero either—this certainly looks as if there is some kind of regulation. The fundamental principle of natural selection is that over-production of young causes competition. Following this principle, most interpretations of fluctuations in population numbers involve competition for resources, mainly food. Hence theorists concentrated first on how resource competition could be related to density.

Explanations of population dynamics were therefore sought from the effects of density on reproduction and survival, although, as mentioned before, there were always some dissenters who thought that density-independent factors such as 'weather' could regulate populations. For most of the twentieth century, ecologists concentrated on understanding the impact of density on birth and deaths because they assumed that populations were closed, hence movement did not happen or was insignificant. If we can assume a closed population, then we only have to consider births and deaths. These models dominated ecological thinking for a very long time.

A nice example is seen in an 11-year study of the American Redstart *Setophaga ruticilla* (Figure 2.5a) These birds need territories to breed. Using the number of territories per 100 hectares as a measure of density, we can see the effects of density in two different measures of reproductive success (Figure 2.5b,c). Both of these measures are combinations of reproduction (the number of eggs laid) and mortality (the death of young birds before fledging). We can also see the impact on the rate of population increase (Figure 2.5d). In Figure 2.5d, notice the scale of the y-axis. The growth rate of the population can be negative or positive, but is a function of density with a growth rate of zero at a density of about 38 territories per 100 hectares—this has a special term in ecology, called the carrying capacity of the environment.

The argument about density dependence hinges on the relationship seen in Figure 2.5d. In this case, there is a straight line connecting density with the population growth rate. Even small changes in density cause changes to the growth rate, and these changes lead to increases or declines in density that keep the average density at the carrying capacity of the environment.

What about the thrips of Figure 2.4d? The numbers per rose constantly fluctuate, but does that mean thrips have no carrying capacity? Based on such data, some entomologists think that insect numbers are generally well below the carrying capacity except perhaps on rare occasions. They interpret this to mean that density-dependent mechanisms rarely operate in insect populations, and hence that insect populations are not regulated by them. An alternative explanation is that the scale of the study is simply not large enough, just like in the case of the lynx where only a Canada-wide view showed what was really happening.

The *slope* of the relationship of Figure 2.5d is also probably important. It describes by how much the growth rate changes for a given change in density from the carrying capacity. If the slope is steep, then the response is strong, whereas if the slope is shallow, then the response is weak. It turns out that differences in this slope are all that are needed for very simple theory to explain

Figure 2.5 a–d Successful breeding in the American Redstart (*Setophaga ruticilla*) leads to an equilibrium number of territories in a given area. (a) a male American Redstart; (b) the number of offspring successfully fledged per mother plotted against density (the number of territories per 100 hectares); vertical lines represent the standard error of the mean for each of the years, and the solid line is the fitted line of best fit (the regression line); (c) relationship between density and the proportion of mothers that successfully fledge at least one offspring; the line of best fit is plotted; (d) the relationship between the rate of population growth and density; the line of best fit (solid) is plotted, and the horizontal dotted line shows the position of zero population growth rate.

(a) Gerald A. DeBoer/Shutterstock.com, (b) Reproduced with permission from McKellar, Ann & Marra, Peter & Boag, Peter & Ratcliffe, Laurene. (2014). Form, function and consequences of density dependence in a long-distance migratory bird. October 1, 2013. *Oikos*. Copyright © 2013, John Wiley and Sons.

all the patterns of population dynamics seen in Figure 2.4. Weak slopes mean that the population dynamics are like the sheep (Figure 2.4a), whereas steep slopes produce the apparently chaotic dynamics of the thrips (Figure 2.4d).

Patchiness and movement

The theory of a metapopulation, a 'population of populations', was worked out in 1969 by Richard Levins, but as with many other major ideas, its significance lay unappreciated for a decade or two. It came back into use in large part because of another important work—one of the most influential books ever published in ecology. This was '*The Theory of Island Biogeography*' by Robert MacArthur and Edward Wilson, published in 1967. The book introduced more than one new concept to ecology, but here we concentrate on the idea of the dynamic equilibrium in species richness.

The species richness of an area means the number of species that exist there. MacArthur and Wilson were thinking about islands, but the idea is relevant to any environment perceived as a patch for the populations of organisms that live there. In ecological terms, a patch is a discrete area different from its surroundings that provides a population with the resources it needs to survive there for at least a few generations. Natural landscapes consist of many patches of suitable habitat in a sea of unsuitable habitat, just like real islands. The full life cycle needs to occur. Thus a robin flying into an apple tree is in some sense 'immigrating' to a 'patch', but no-one would think it had established a population there that went extinct when the robin flew out again.

The new theory was invented to explain the species richness of islands, and in particular the fact that there are more species on larger islands. If you were to ask any ecologist why this might be, the first idea they would come up with would be that larger islands have more habitat types, and since each habitat type contains a set of species adapted to living in it, then inevitably larger areas contain more species. And, of course, this is true.

The equilibrium theory of island biogeography

Is there anything else that might explain the observed island effect? Edward Wilson's PhD student, Dan Simberloff, carried out some amazing experiments on small mangrove islets, each only a few hundred square metres, off the coast of Florida. He showed that you still get more species of tree-living arthropods on larger islands, even though there is only one habitat—the mangrove.

He showed this by carrying out the gold standard in science—a manipulation experiment. Instead of changing island size experimentally, he could have merely counted the number of species on islands of different sizes. Counting would produce a set of observations, not an experiment. The difference between these methods is a crucial one. If you only observe, you can never conclude anything about causes. So for example, there is an excellent statistically significant correlation between the human birth rate in Denmark each year during the twentieth century and the density of breeding storks. But does that mean this is a causal relationship—do storks bring babies? Of course not! We would need a manipulation experiment to reveal a causal link, experimentally increasing or reducing the density of storks in selected areas, and recording the resulting human birth rates. The correlation arises because both human birth rate and stork breeding are both responding to an unmeasured unknown third factor (probably human-induced changes in land-use).

Simberloff took drastic actions to perform his manipulative experiment. Having counted the initial number of arthropod species, he simply chopped some of the islands in half and got rid of one of the halves! This was just about possible for mangrove islets because they were small, and had little or no land surface. He therefore experimentally changed the area without altering the number of habitats. There were also controls that were not manipulated, and some islands were reduced in area a second time. With a year in between each manipulation, the result was very clear: the number of species reduced every time the area was reduced. Thus we can conclude that on top of the effect of the number of habitats, there was a separate, independent effect of habitat area.

MacArthur and Wilson speculated that this area effect might be due to a balance between two processes—colonization of the island by new species from a 'mainland', and extinction of populations already on the island (Figure 2.6a). The rate of colonization was assumed to be set by the distance of the island from the mainland. This was a reasonable assumption and actually a well-known effect: isolated islands have very few species on them relative to less-isolated islands. The rate of extinction of populations on an island was assumed to be related to its area. This was also reasonable, and supported by a lot of evidence. The reasoning was that the area of the island determined the total number and amounts of resources to be shared by the organisms living there. Smaller islands had fewer resources which could therefore only support smaller populations, and small populations were much more vulnerable to the random effects of bad weather, fluctuating resources, and simply bad luck. Thus the species richness of an island was suggested to be an equilibrium between the extinction of species on the island, and the colonization of the island by new species (Figure 2.6b).

Again, Dan Simberloff performed some extraordinary experiments to show that this was actually the case. He erected huge plastic bags over selected mangrove islets, and used poison gas to kill all the invertebrates. This drastic step was required to bring about the breakthrough in understanding that followed. Over the next five years, he watched the recolonization process and saw that it re-established a very similar number of species as occurred before the treatment. There were some extinctions of populations, and there were colonizations, and together after five years these rates were in approximate balance, the *same balance* that had existed before the experiment began (see Figure 2.6c). The theory was vindicated: there really did appear to be an equilibrium between colonization and extinction.

These startling results came from some of the most innovative and ambitious field ecological experiments ever undertaken. Their influence and ramifications have been profound in ecology and conservation. One of their consequences was to reorientate the focus of ecologists from studies of the regulation of small closed populations onto studies at a much large scale that included movement and population extinctions. And out of that reorientation came support for the theory of the metapopulation—the cornerstone idea of modern conservation science.

Metapopulations

Across a set of patches or islands, the idea behind the equilibrium theory of island biogeography is very similar to the idea behind the metapopulation, except in the first we are concerned with species richness, while the second involves a set of populations of one species. Both are concerned with

Figure 2.6 a–c The ideas behind the equilibrium theory of island biogeography: (a) ideal thought experiment—take a set of new islands with nothing on them, that differ only in area and in distance from the mainland (the source of colonists); (b) the rate of colonization of new species onto an island continues until everything that can colonize has done so (the height of the curve depends on distance alone); the rate of extinction increases as the populations of more species share the limited resources of the island, a limit set by island area alone; (c) the results of Simberloff's experiment—the recolonization of mangrove islands experimentally wiped clean of species re-established the same equilibrium number of species as occurred there before.

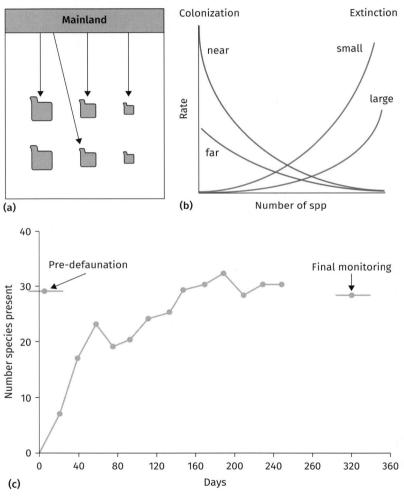

movements of individuals among patches (or islands), and with extinctions of populations on patches (or islands). Like the equilibrium theory, the metapopulation represents a dynamic balance between immigration and extinction.

In Richard Levins' original idea of a metapopulation, all patches were identical and distances among adjacent patches were the same—it was a theoretical model of identical elements. Although we know that biological systems are never identical, this type of model can be very useful. Like previous population models it was a closed system, but at a much larger scale than that of a single patch.

Levins' model investigated what happened when the population of a single patch was not viable on its own. Inevitably, left to itself, this population might fluctuate up and down either in response to annual weather changes, or simply to the random events of demography (such as the sex of the offspring and the decision to emigrate). In a very small population, these can be critical: if the last female only has male offspring then the population is inevitably doomed. The question Levins set himself is this: what happens if you have a set of vulnerable populations with a small amount of movement in between the patches?

What Levins discovered was important, fascinating, and alarming all at the same time. If the rate of immigration exceeded the rate of extinction, then on average the metapopulation was perfectly stable through time, and hence never died out. Each individual patch population might die out at any one time, but it could be rescued from dying out by immigration of new individuals at the critical time, or it could be re-established by immigration at some future time. Individual patches might therefore wink on or off, but overall the metapopulation kept going. If the rate of immigration was reduced to be equal to the rate of extinction, this stability was still the end result. However, if the immigration rate dropped below the extinction rate, then the entire metapopulation suddenly died out!

Of course, real patches and populations are not all identical. Patches vary a lot in size and in isolation from one another, and patch populations also vary a lot in numbers, composition, and the degree of synchrony in their fluctuating numbers. However, the world is definitely patchy at lots of different scales. And, crucially, human use of the landscape is creating lots more small patches out of previously continuous habitat. These ideas were therefore rapidly seen to be highly relevant to conservation.

After a lot of fiddling with different models, and a lot of experimental and observational studies in the field, it gradually became clear that differences in the extent of movement, and in patch sizes and distance, really matter to the outcome. Where you have lots of movement, then even though the environment looks patchy to us, to the organisms it clearly isn't and the situation is best treated as a single population. When you have no or too little movement, then emigration and immigration do not matter enough to be able to rescue patch populations from extinction, or to re-establish populations that have died out. Here there is no metapopulation, but merely a set of separate populations on the patches. When you have a single large 'mainland' patch with a population large enough so that there is zero risk of extinction, then again no true metapopulation is formed, and instead an 'island–mainland' scenario is the right one. It is only when you have a medium amount of movement, that rescues or re-establishes extinction-prone populations, that a true metapopulation structure exists.

Case study 2.1
The Glanville Fritillary *Melitaea cinxia* (Lepidoptera: Nymphalidae)

The best-known study of a metapopulation is that of the Glanville Fritillary (*Melitaea cinxia*) in the Åland Islands in Finland, by Ilkka Hanski and his group over the last 25 years. This butterfly lives in dry meadows, which form more than 4000 discrete habitat patches across the one large, several medium-sized, and many small islands in the archipelago. The occurrence of either of the two larval host-plants (Ribwort Plantain, *Plantago lanceolata* and Speedwell, *Veronica spicata*) is critical for the presence of the butterfly. Because the larvae feed gregariously in groups of 50–250, there must be enough host-plants for the group to be able to move to a new plant when the old one has been eaten. Patches are better if grazing keeps the grass down, helping the host-plants to avoid being smothered.

The meadows have been surveyed every year since 1993, and actually form 125 more-or-less separate metapopulations, divided into 33 viable and 92 inviable ones (see Figure A). In the inviable networks, the patch sizes, qualities, and degrees of isolation all hinder butterfly survival, and there were no butterflies in 79 of them for at least five of the 22 years surveyed. In the viable networks, the nature of the patches helps butterfly survival, and here only five of the 33 networks were extinct for at least five years. In the short term, the inviable networks can act as temporary stepping stones to help the species to spread across large areas. Such 'corridors' that improve the prospects

CS2.1 Figure A The 125 separate metapopulations (polygons) of the Glanville Fritillary (*Melitaea cinxia*) on the Åland islands of Finland, consisting of a set of viable metapopulations (red, consistently occupied) and a set of inviable metapopulations (light-blue, usually unoccupied). The scale bar is 10 km.

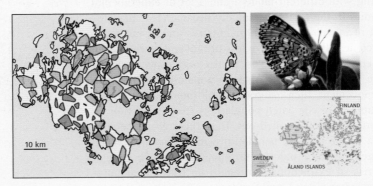

Adapted from Hanski, I. et al. Ecological and genetic basis of metapopulation persistence of the Glanville fritillary butterfly in fragmented landscapes. Nature. 2017. Distributed under the terms of the Creative Commons Attribution 4.0 International (CC BY 4.0). https://creativecommons.org/licenses/by/4.0/. Photograph Boza C/Shutterstock.com.

for organisms moving from patch to patch turn out to be very important elements of the landscape in conservation planning.

As with the Large Blue butterfly (see Chapter 1), there are some genetic effects that influence the metapopulation, and once again these are connected with dispersal ability. In this case, the effects have been connected to a single-nucleotide polymorphism of one gene, the glucose-6-phosphate enzyme (Pgi) of glycolysis. CC homozygotes and AC heterozygotes have a higher flight metabolism and are better at moving between patches than the AA homozygotes. This increases the rate of colonization, and decreases the rate of extinction by rescuing populations more frequently, and thus shifts the likelihood of metapopulation persistence. This is a very rare example of how a single base change in DNA can influence landscape-level processes.

 Pause for thought

Do you think snails are likely to exist in metapopulations? What is the important factor(s) in reaching your conclusion?

 Chapter summary

- Understanding the dynamics of populations involves studying the key demographic processes of birth, death, immigration, and emigration. Much of twentieth-century population ecology concentrated on explaining population dynamics of closed populations, using the effects of density on births and deaths, ignoring movement.
- The equilibrium theory of island biogeography changed this focus onto the study of movement as a key determinant of species richness.
- This focus was transferred into the idea of the metapopulation, a set of extinction-prone populations on patches kept going by the movement of individuals.
- Natural landscapes consist of many patches of suitable habitat in a sea of unsuitable habitat.
- Movement across such landscapes is an important element in the population dynamics of many organisms.

 Further reading

Levin SA (ed) *Princeton Guide to Ecology*. Princeton UP. pp. 431–437.

The Princeton guide consists of short (4–8 pages) chapters written by different authors outlining concepts and ideas in ecology and conservation. Some are rather detailed, but they expand and consolidate the ideas presented here.

Discussion questions

2.1 If a population on a patch depends on immigration from outside for its survival, what are the implications for nature reserves?

2.2 Is it worth having small nature reserves at all?

2.3 If the gold standard for science is a manipulative experiment, where does this leave much of medical science?

3 RARITY AND EXTINCTION

This chapter considers the consequences of changes in the size of a patch of suitable habitat. The key issue to understand is the vulnerability of small populations to dying out simply by bad luck.

The changes to the landscape made by twenty-first-century humans are overwhelmingly to reduce the amount of natural habitat and to increase its fragmentation, replacing natural with agricultural and urban habitats. Until a

Figure 3.1 Cape fox *Vulpes chama* in the Kalahari desert—a vulnerable species in a special habitat threatened by human-induced changes. Are nature reserves the answer?

© Andrew MacColl

few years ago, conservation action consisted of advocating the setting aside of nature reserves (Figure 3.1). However, in the end this results in isolated islands of natural habitat in a sea of agricultural or urban land. We explore the likely consequences here.

Rarity and extinction

Both the equilibrium theory and the theory of metapopulations directed the attention of population ecologists towards larger scales, and to movement and population extinction as being critical factors. What affects the chance that a population will become extinct?

Once again, the early work was done on islands in the context of the equilibrium theory of island biogeography, only to be superseded by experiments involving metapopulations.

Look at Figure 3.2a, which comes from work on the birds of New Guinea by Jared Diamond. Note that during the last Ice Age, a great deal of seawater was locked up in the ice-caps, hence the level of the sea was lower relative to today. This means that a lot of land was exposed then, but is now under water. Thus some areas that are islands today were then parts of the mainland—these are what Diamond calls 'land-bridge' islands.

Now study Figure 3.2b, which shows a plot of the species richness of the birds of the islands around New Guinea (the solid circles), together with the number of bird species measured on equivalent areas within the mainland of New Guinea (the crosses). The axes are not straight numbers but powers of ten (i.e. logarithms), but this is not important in the interpretation of this remarkably informative graph. Diamond developed all sorts of hypotheses from it.

The first thing to notice is that there is a positive relationship (the red line) between island area and the number of species for the 'oceanic' islands. The second is that there is also a clear relationship (the blue line) between area and species richness for the sampled areas on the mainland of New Guinea. So far so good—nothing remarkable there. The important issue is that these two relationships are not the same. The gradient of the line for the mainland is higher and shallower than the line for oceanic islands, hence the gap between them widens as area gets smaller (to the left). Why should this be? Oceanic islands have many fewer bird species than an equivalent area on the mainland, and smaller islands suffer more from this deficit than larger islands.

Diamond used the data for the land-bridge islands (the open circles) to infer the process that was happening. The land-bridge islands all lie between the lines for the mainland and the oceanic islands. Thus, he

Figure 3.2 (a) New Guinea and surrounding islands. The shaded islands ('land-bridge' islands) were part of the mainland during the last Ice Age when the sea level was much lower than now - shown by the dotted lines. The unshaded 'oceanic' islands outside the dotted line have always been islands. They have never been part of the mainland. The bird is a Wilson's Bird-of-paradise *Diphyllodes respublic*. (b) the number of bird species (*y*-axis) against area (*x*-axis) for areas of the mainland of New Guinea, land-bridge islands, and 'oceanic' islands); the trajectory in terms of species richness through time of one land-bridge island (Misol) is shown by the dark blue arrow (what has happened since the last Ice Age) and the light blue arrow (what is yet to come).

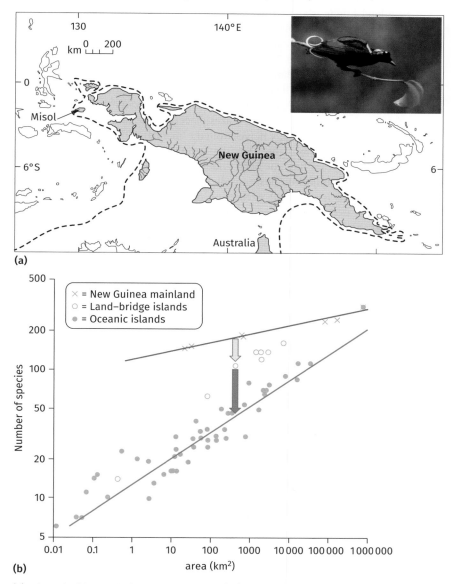

(a)

(b)

(a) Adapted with permission from Nitecki, M. (ed) *Extinctions*. University of Chicago Press. Copyright © 1984. (ai) AGAMI Photo Agency/Alamy Stock Photo (b) Adapted with permission from Diamond, J. M. & Nitecki, M. H. (Ed). "Normal" extinctions of isolated populations. In *Extinctions* (1984), pp. 191-246. University of Chicago Press.

reasoned, when they were part of the mainland, the species richness of their area would have been part of the 'mainland' line and hence would have been much higher. Then about 10 000 years ago the Ice Age ended, the ice caps melted and the sea level rose, cutting them off as islands from the mainland. Now, their species richness 'should' belong to the island line, and therefore is much too high. Each land-bridge island then loses species as its position on the graph gradually subsides down from the 'mainland' to the 'island' line.

Why does this happen? In terms of the MacArthur and Wilson theory, the new land-bridge island used to be part of a much larger area, but now the area has contracted. This means there are fewer resources to share amongst all those bird species, hence the populations become much smaller and more vulnerable to extinction. Furthermore, the new land-bridge islands are now a distance from the mainland, and hence much harder to move to than before. The extinction rate has therefore risen, and the immigration rate dropped, implying a much lower equilibrium number of species. As the excess number of extinctions over immigrations takes place, the position of the land-bridge island on the plot gradually moves from the 'mainland' line to the 'island' line.

As is obvious from the plot, the extinctions do not happen quickly. Even 10 000 years has not been enough to move the land-bridge islands to their true equilibrium positions dictated by their new area and degree of isolation. This is called the extinction debt, the number of species that are doomed to extinction in the long run, but are still hanging on. Note that the smaller the land-bridge island, the more complete is the process of loss (i.e. the closer to the 'island' red line).

Although this example uses data from birds, in fact the process seems very general and applied to many different kinds of taxa.

Scientific approach 3.1
Moss modelling

The work by Jared Diamond with birds is an observational study: can we match this with an experimental study, so as to be sure about the causal effects of changing area? Well, we certainly cannot experiment with processes that last 10 000 years, so the only option is to use a micro-ecosystem. This works because smaller organisms reproduce and die more quickly—one of the most general of biological laws. Some bacteria can reproduce by dividing every twenty minutes; *Drosophila* flies take a maximum of a couple of weeks to develop from the egg to the egg-laying adult female. Thus if you can miniaturize the system, you stand a chance of seeing the process of population extinction stimulated by area contraction because all the demographic process (births, deaths, movements) happen so much more quickly. There is a long tradition—more than a century—of using such 'microcosm' experiments in ecology.

SA 3.1 Figure A A carpet of moss on stones is home to a variety of small organisms—it is an entire miniature ecosystem.

Mr Doomits/Shutterstock

This is exactly what was done as an undergraduate project at Nottingham University. Students used the small organisms that live in moss—mostly different species of mites and springtails—as their model ecosystem (Figure A).

Students took expanses of moss on large flat limestone rocks under trees, and cut away the moss to create small patches, 'islands', exactly as the sea had done around New Guinea. Each island was just 10 cm in diameter, and at least 10 cm from other areas of moss, separated by bare rock presumed to be hostile to the movement of mites and springtails because of the risk of dying from desiccation. After six months, judged to be long enough for extinctions to happen in the patches, each island was sampled in a very simple way by taking it off the rock and putting it in a **Tullgren funnel** for 24 hours. This is just a vertical tube with a light bulb at the top to generate a bit of heat, and a jar of alcohol at the bottom—the heat dries out the moss, and the animals try to escape this by moving downwards, causing them to drop out into the alcohol. Having the specimens enables them to be identified.

The results are shown in Figure B—the experiment was designed to test two ideas:

- Does reducing the area reduce the number of species, after a suitable time has elapsed for the extinctions to happen?

- Can we help prevent some of these extinctions by encouraging movements among the islands through 'corridors'—strips of moss running between the islands? From the previous discussion about metapopulations, you can see that movements could rescue populations from dying out on any one island.

SA 3.1 Figure B Experimental fragmentation using moss 'islands', showing the average number of animal species (mostly mites and springtails) remaining in each 10-cm 'island' (lower line), or the set of four islands collectively (upper line), six months after fragmentation. The error bars are standard errors of the mean. There were four replicates of each treatment on each of six large flat limestone rocks.

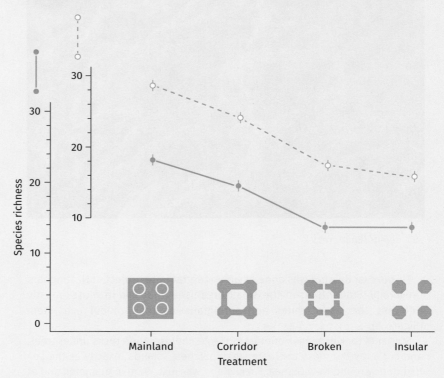

Reproduced with permission from Gilbert, F. et al. Corridors maintain species richness in the fragmented landscapes of a microecosystem. *Proceedings of the Royal Society B (Biological Sciences)* 265 pp. 577-582. Copyright © 1998, DOI: 10.1098/rspb.1998.0333

Thus there were control 'islands' cut out of moss carpets, two sorts of 'corridor' treatments (entire and 'broken' corridors—the latter to prevent movement but to control for the extra area of the corridors), and isolated islands.

There turned out to be about 100 individual animals of between 8 and 25 species in each island, and the pattern of the number of species among the treatments was very clear (see Figure B). There were about 23 species in an island cut from a continuous carpet of moss, but only about 14 in an isolated island. Crucially, notice the difference between the treatment with continuous corridors that allow movement (with about 20 species per island) and the one with broken corridors that prevent movement (with about 14 species, the same as the isolated island). Thus enabling movement prevents most of the extinctions that happen in isolated islands.

Why are such experiments relevant to the theme of this book? The answer lies all around us. The overwhelming change that has happened to the landscape because of human beings is habitat fragmentation and habitat loss. An example is shown in Figure 3.3. Loss and fragmentation are separate but connected issues; you can have habitat loss without fragmentation, but usually they go hand in hand.

This means that the main change happening to the natural world is that islands are being created from previously continuous habitats, and these islands are gradually getting smaller and more isolated. From the discussion we have had about the relationship between area, isolation, and species richness, you can immediately see that this creates a massive problem. There is a huge extinction debt building up as a consequence of what we have already done. For many plants and animals, the populations are still here but they are effectively doomed because they are too small to resist the fluctuations that happen in the long term, and eventually they will die out if habitat availability does not improve.

It is easy to think that all the environmental damage has been done in the last few years—but a look at the forests of eastern Germany (Figure 3.3) shows that our impact began much earlier than that. 92.4% of the forest existing in 1780 had been lost by 1880, and a further 1.8% by 2008. This was slightly offset by an 8.5% gain through replanting, but such forests do not contain the rich understorey flora of ancient forests. While much of the habitat loss occurred early, most of the fragmentation occurred between 1880 and 2008: the number of patches increased from 1075 in 1780 to 1459 in 1880 and a remarkable 5964 in 2008. Most of the understorey plants that disappeared did so fairly quickly—the only remaining evidence for an extinction debt is among forest specialist herbs.

This example comes from Germany rather than the UK because there is no equivalent study for the UK. We lost most of our forests even earlier, before the existence of adequate documentation, and now (along with Ireland and the Netherlands) we have the lowest proportion of woodland of all the countries of Europe.

Human beings can therefore create metapopulations where none existed before, by reducing habitable areas and increasing the degree of fragmentation, making movements more difficult and populations more precarious. A good example of this is in tiger populations in India, which are seriously in decline. Modern tigers show all the genetic hallmarks of a fragmented population, with an increase in the genetic differences among populations relative to those existing in tigers before 1950.

What can we do?

You might imagine that setting aside nature reserves is the answer to this conservation problem. It can help, but by itself it is not going to work. A famous study from 1987 by William Newmark showed this very clearly. He studied the mammals of western North American National Parks and park assemblages, ranging in size from Bryce Canyon (144 km²) to Grand Teton–Yellowstone (10 328 km²) and the huge Banff-Jasper (20 736 km²).

Figure 3.3 Habitat loss and fragmentation of ancient forests in the Prignitz region of eastern Germany.

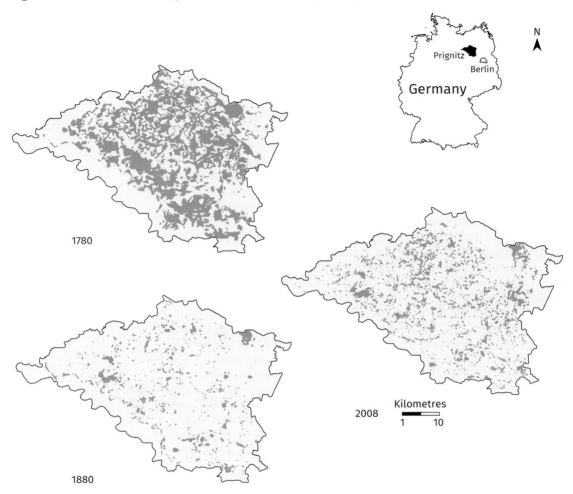

Some of these parks share borders, giving an enormous protected area. Each park was well funded and well managed, so their mammal populations were known in some detail, especially the relatively large and popular species such as rabbits, hares, carnivores, deer, etc. that Newmark worked with.

All the parks had been in existence for more than 60 years, so Newmark was able to consult the records back through to the establishment of each park. He counted all the species that had been there and had disappeared since establishment, together with all those that had colonized each park since establishment. He then looked in detail at the causes of each colonization or extinction, to separate obvious human-influenced change from 'natural' changes—i.e. those for which there was no obvious cause. Park managers had reintroduced 12 species found historically there, and four exotics ('alien' invaders to North America) had colonized, but there were only three natural colonizations. These had followed natural range expansions of the Raccoon (two parks) and the Moose. In contrast, there had been 42 natural extinctions after the establishment of the parks, accompanied by a further 42 pre-establishment and six human-influenced extinctions. Thus extinctions vastly outweighed colonizations across all of the parks.

This is exactly what we might expect if we think about parks and reserves as islands of natural habitat in a sea of agricultural landscapes. They are or rapidly become smaller bits of natural habitats than before, and also become more isolated, exactly as the land-bridge islands that Diamond studied. Thus we predict that the rate of extinction should rise, and the rate of colonization should fall, and overall we should see a loss of species with time. The losses should be related to 'island' area, and this is exactly what Newmark found for North American parks (Figure 3.4b). Only the largest park assemblage, the Banff-Jasper system, still contains an intact set of these mammals, the same set that it started off with; Olympic Park also had zero natural extinctions, but lost one species due to human-induced factors. Furthermore, there is even a relationship with the age of the park: for their size, parks lose more species the older they are. Both of these relationships, with area and age, are exactly what we would expect from MacArthur and Wilson's theory of island biogeography.

Figure 3.4 (a) The number of natural (i.e. not human-influenced) extinctions of mammals in North American parks that occurred after the establishment of the park, plotted against the area of the park. The line of best fit suggests that only the very largest park retains its original mammal assemblage. (From Newmark 1987 *Nature* 325: 430). The photos (bi and bii) show a Bison (*Bison bison*) and a Rocky Mountain Goat (*Oreamnos americanus*).

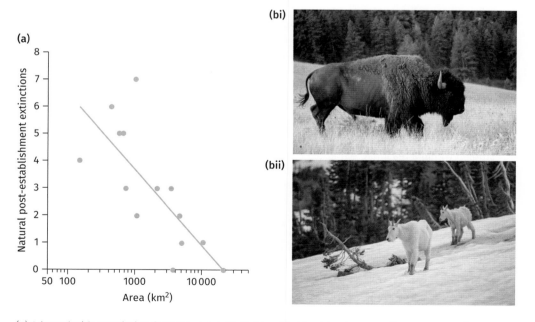

(a) Adapted with permission from Newmark, W. D. A land-bridge island perspective on mammalian extinctions in western North American parks. *Nature*. Jan 29, 1987. Copyright © 1987, Springer Nature. [https://doi.org/10.1038/325430a0] (bi) Tory Kallman/Shutterstock.com (bii) Steve Boice/Shutterstock.com

For conservation biologists these results came as a shock. They meant that even the huge parks of North America were not large enough to prevent mammal extinctions caused by the combination of area loss and increasing isolation. They emphasized that conservation was operating at the wrong scale, and we needed to think very differently and on a much larger scale.

The inadequacies of European and particularly UK parks and reserves were obvious. We cannot easily increase the areas of parks without some painful decisions about depriving landowners of their land, or removing human villages and towns, so the next-best strategy is to connect them up with corridors of land so as to boost movements between parks.

These are exactly the measures called for by the Lawton report of 2010. Professor Sir John Lawton is an ecologist asked by the UK Government to review what could be done in the UK to improve nature conservation in a sustainable way. The title of his report is 'Making Space for Nature', and throughout it is concerned with issues of area and connectivity—just the ideas we have been exploring here. In it he argued that the UK needs a step-change in its approach to nature conservation. Instead of just trying to hang on to what we have in the face of the various competing demands for land, we need to embark on large-scale habitat restoration: not just connecting up the existing reserves into a network, but increasing the number and extent of the places where conservation and sustainability are first in the queue where land-use is concerned.

Case study 3.1
The Sinai Baton Blue butterfly *Pseudophilotes sinaicus* (Lepidoptera: Lycaenidae)

Butterflies have been used extensively to study metapopulation processes because they can be marked and observed easily (many studies show that marking does no harm to butterflies). Their population changes can then be studied within a reasonable timescale. In this case study we introduce South Sinai, where we have researched biodiversity for more than three decades. Many of the case studies in this book are taken from our work there.

South Sinai is a very special place, ecologically, historically and culturally. The map (Figure A) shows it as a semi-detached region of Egypt—the cradle of one of the world's oldest civilizations, and officially the driest country in the world. In 1996 nearly all of the mountain massif of South Sinai was declared as the St Katherine Protected Area (Figure B), about half of the entire administrative region. Protected Areas with shared boundaries to the east and south mean that almost all of the southern half of the peninsula is now part of a protected reserve. The availability of water drives all life there.

At the heart of South Sinai lies the great Ring Dyke, a huge incomplete black ring of volcanic rock surrounding a red granite core. In its centre is the town of St Katherine, the only Bedouin-majority town in Egypt, and the only town inside the whole park. The town is relatively recent: before the Israeli invasion during the 1967 war there was no town, merely a cluster of Bedouin houses near the Monastery of St Catherine, the world's oldest monastery in continuous occupation since its establishment in the year 560 CE.

Egypt was not always so dry. After the end of the last Ice Age, around 7500 years ago, it had a much cooler and wetter climate and animals including spotted hyenas, warthogs, zebra, wildebeest, and water buffalo lived there. In the early days of the first Pharoahs, about 4500 years ago, conditions became more arid but there was still a rich mammal fauna including lions, wild dogs, oryx, hartebeest, and giraffe, none of which exist in Egypt today. We know that they were present from their artistic representations on tomb walls, and on pottery and other items made by the Ancient Egyptians.

As the whole environment became hotter and drier, so the fauna and flora had to respond. They could either evolve to cope with these hotter and drier conditions (unlikely); or, if they were capable, move north to track the conditions they preferred; or, finally, they could move up in altitude, since this also allows species to track the conditions that they prefer. Higher altitudes were available in South Sinai, and as a result many species became marooned on these mountaintop islands as other elements of the same communities moved north. Thus the nearest similar communities to those of the South Sinai mountains are hundreds or thousands of kilometres further north.

CS3.1 Figure A Shows the region of South Sinai and its size and area, together with the area of the St Katherine Protectorate (StKP) enclosed within the orange arrows.

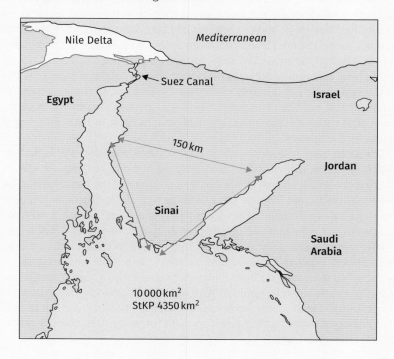

CS3.1 Figure B The St Katherine Protectorate (solid black line) is a park enclosing most of the mountain massif; the only tarmac roads are shown as dotted lines; the only towns are indicated.

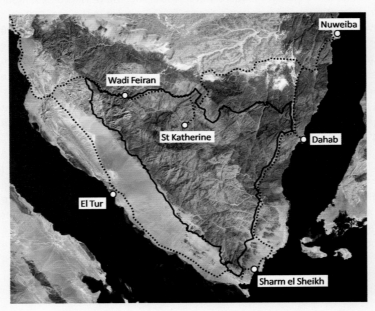

There are many species currently thought to be endemic to the mountains of South Sinai—they are not found anywhere else in the world, as far as we know. They include at least 33 species of plant, but the number of animal species, especially insects, has not been properly assessed—at the moment only two butterflies are known definitely to be on the list.

The Sinai Baton Blue butterfly (*Pseudophilotes sinaicus*) (Figure C) is one of the two species of butterfly thought to be endemic to the mountains of South Sinai. It was first collected in May 1974 in the high mountains above the town of St Katherine. Since the return of Sinai to Egypt in 1982, no biologist had seen this species.

The Baton Blues (*Pseudophilotes* spp.) make up a small genus of butterflies, with only nine described species worldwide. All of them are mountain species of dry rocky slopes, all with larvae that feed on plants of the family Lamiaceae, especially the thymes (*Thymus* spp.). Until the turn of the millennium, none of the species was well studied. The upperside of the wings of the male Sinai Baton Blue is brilliant iridescent blue (Figure C), while that of the female is dark brown.

Mike James, a PhD student, went to Sinai in 2000 to study the Baton Blue, not knowing whether it still existed, or how to find it. He was delighted to see a specimen emerging from under its host-plant—a gorgeous male—but it was promptly eaten by a spider! Mike worked with an expert local Bedouin guide, Farhan Zidan, under extremely difficult conditions of heat (up to 38°C) and aridity (all water had to be carried up to the study sites, 600 m higher in altitude than the town).

CS3.1 Figure C The Sinai Baton Blue butterfly, photographed by the scientist who rediscovered it, Dr Mike James. It is probably the smallest butterfly in the world. The larva is feeding on the flowers of its only known host-plant, Sinai Thyme—note the ant riding on its back.

Images © Mike James, by kind permission

The Sinai Baton Blue is totally dependent on a single plant, Sinai Thyme (*Thymus decussatus*), found only in the South Sinai mountains and a small area of Saudi Arabia. The butterfly larvae only feed on thyme buds and flowers, and the adults feed mainly but not exclusively on the nectar of the flowers. Sinai Thyme is very patchily distributed, growing as more or less widely separated individual bushes above 1500 m altitude in suitable soil of rocky narrow gorges and also in open areas bordered by steep bare cliffs. After exhausting climbs all over the mountains, Mike found 41 patches, but only 25 of them were occupied by butterfly populations (see Figure D); subsequent studies by another PhD student, Katy Thompson, discovered more patches of thyme, so the total number is probably between seventy and one hundred.

The entire world population of the Sinai Baton Blue is apparently contained within a circle of radius 7 km, and in 2001 the total population was estimated at below 3000 adults. Flowering of the thyme depends on weather in a complex manner, and thyme plants can be killed by prolonged drought—thus the food of the butterfly larvae varies a lot from year to year. The degree of synchrony between butterfly and plant flowering varies as well. The consequences are that butterfly populations can change greatly between years, but surprisingly these changes seem to form a three-year cycle whose mechanism is not currently understood. The low points can be very low, hence the risk of extinction for any one population can be high. All these factors combine to classify this butterfly as Critically Endangered on the International Union for Conservation of Nature's (IUCN) Red List scale of extinction risk.

CS3.1 Figure D The entire world distribution of the Sinai Baton Blue butterfly, *Pseudophilotes sinaicus*. Yellow arrows mark the largest and the most important populations. Patches of its host-plant the Sinai Thyme *(Thymus decussatus)* are marked: yellow patches contained butterfly populations in 2001 and 2003; grey patches had populations in 2001 but not in 2003; white patches had no evidence of butterflies in either year. Note the scale: the entire map is only a few kilometres wide.

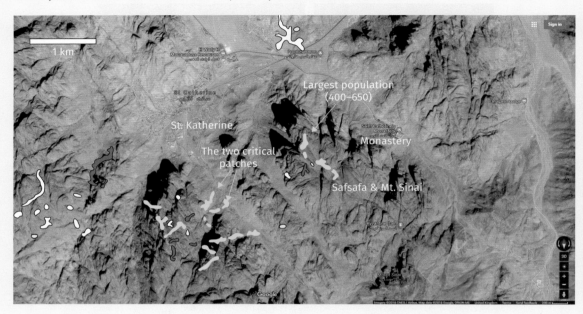

Imagery © 2018 CNES/Airbus, Map data © 2018 Google, ORION-ME.

Life cycle of the Sinai Baton Blue

There is just a single generation of the butterfly per year, and it overwinters as pupae in the soil beneath the host-plant. Newly emerging adults crawl out of the soil then up to the top of the thyme plant, where they remain (often for several hours) until their wings have expanded and dried. They emerge between the end of March and late July, but mostly in May and June. The exact timing varies from year to year, but coincides roughly with the flowering of Sinai Thyme. Males emerge before females in order to mature their reproductive system ready to mate with emerging females: most females probably only mate once, often before they take their first flight. Mating lasts for over an hour!

The rest of their short adult lives is spent flying about, feeding, laying eggs on thyme buds (females), or patrolling 'territories' and mating (males). Both sexes try to avoid being eaten by spiders, praying mantids, birds (especially Scrub Warblers *Scotocerca inquieta)*, and particularly lizards: predation is probably the main source of mortality. The adult lifespan is normally very short, only 3–6 days.

Typically the eggs hatch after four to five days. There are five stages ('instars') of the green larvae during their three-week development. First instars are really tiny and cannot move far from the egg, feeding only on the buds immediately at hand. Fully developed fifth instars are still only 8–9 mm long, and are still probably restricted to the plant on which they hatched. They are 'tended' by ants (Figure C) but, unlike the Large Blue (*Maculinea arion*), these larvae do not require ants in order to complete their development. The tending ants belong to the genus *Lepisiota*—but there are other ants in the genus *Crematogaster* that eat butterfly larvae. *Lepisiota* nests are found in the rockier parts of the habitat, and *Crematogaster* ants in the more open areas. As a result, none of the butterfly eggs laid on plants in open areas produce adult butterflies—they all get eaten. In contrast, a proportion of the eggs laid in plants in the rocky areas survive to produce adults.

At the end of larval development, i.e. from the end of May onwards, the mature larva climbs down the plant to the soil, and digs itself down a few centimetres into the soil to pupate. It is in this stage that the Baton Blue survives the rest of summer, autumn, winter, and early spring—when temperatures regularly fall below zero and snow is common. As in related species, a proportion of pupae may take two, three, or even more years before they emerge, a kind of insurance policy against the regular occurrence of years of drought.

Every year, some individuals emerge early before thyme is flowering, and some individuals emerge late, after the end of flowering: both of these groups find that there are no buds on which to lay their eggs, so these individuals tend to emigrate to other patches. Overall this happens to about 12% of individuals.

The network of populations

Mike James surveyed the network of patches of thyme for butterflies in 2001 and 2003 (Figure D). In 2001, patches to the west were mostly unoccupied, as was the large (but very poor-quality) patch of Gebel Sonar in the north. Most of the rest of the network was occupied, but interleaved amongst the occupied patches were at least five unoccupied ones. By 2003, Sinai was well into a prolonged dry spell and no rain at all fell that year—the strongest drought for 50 years according to Bedouin colleagues. Butterfly populations on eight patches disappeared, and hence presumably became extinct (although we do not know about the pupae with delayed emergence). As in many other butterflies and also birds and mammals, the pattern has all the hallmarks of a metapopulation.

Despite the apparent simplicity of the data for the network of patches—presence/absence for two separate years—we can create a sophisticated metapopulation model that incorporates random climatic variation, climate warming, and other risks such as over-collection of thyme by pharmaceutical companies. The model suggests that while the butterfly metapopulation is safe under current conditions, further climate warming or loss of patches from over-collecting will put it at a significantly increased risk of extinction.

❓ Pause for thought

Could you have carried out the fieldwork of this study? In one year, Mike James kept track of every individual butterfly by walking around his study site all day from dawn to dusk every day for 111 consecutive days.

Chapter summary

- Reductions in island or habitat area lead to extinction of some of the populations because of the reduced resources and increased isolation.
- Habitat fragmentation leads to an extinction debt because it takes time for populations to die out.
- Creating corridors between isolated patches can help prevent some extinctions, but they cannot abolish them entirely.
- Setting aside nature reserves for conservation is obviously a good thing to do, but it will not be enough as the reserves become isolated patches of natural habitat in a sea of man-made agricultural or urban landscapes.
- The small areas and increasing isolation of nature reserves lead to extinctions.
- A landscape approach to conservation is needed for it to be successful.

 Further reading

Slobodkin LB (2003) *A citizen's guide to ecology.* **OUP.**

This is a popular paperback written for the general public about ecological ideas. It is a bit out of date in its focus, but contains a lot of useful material presented in an engaging way.

Levin SA (ed) *The Princeton guide to ecology.* **Princeton UP. pp. 438–444, 445–457, 529–537.**

The Princeton guide consists of short (4–8 pages) chapters written by different authors outlining concepts and ideas in ecology and conservation. Some are rather detailed, but they expand and consolidate the ideas presented here.

 Discussion questions

3.1   How might the human-driven climate change of today affect the current distributions of animals and plants in Egypt?

3.2   What do you think might be the most effective way of improving conservation in the UK, using any one of the three elements of improving habitat quality, increasing habitat area, or connecting up habitat patches?

3.3   How important do you think the quality of agricultural land is for wildlife to be successfully sustained?

4 INTERACTIONS AMONG SPECIES

In the late 1970s and 1980s, Dan Simberloff of the University of Florida challenged animal ecologists to produce better evidence for the importance of interactions among species in animal communities. He believed in reductionism, which as we have seen suggests that interpretations at one level of the biological hierarchy (see Figure 4.1) should be sought at the level below.

Figure 4.1 The biological hierarchy.

The earth
Ecosystems
Metacommunities
Communities
Metapopulations
Populations
Individuals
Organs
Cells
Organelles
Molecules
Atoms

Somchai Som/Shutterstock.com

Thus for example, the population changes of one species (the species level) should primarily be studied by looking at what was happening to the individuals in the population (the individual level), rather than the interactions of the species with others in the community (the community level).

The challenge was undoubtedly effective in forcing ecologists to do the relevant manipulation experiments rather than to make assumptions based on simple observations. Since then, a large body of evidence shows that interactions among species are very important in nature. In this chapter you can look at this evidence in a bit of detail.

Herbivores and plants

It took an unimaginable length of time—hundreds of millions of years—for the first living thing to evolve, but it took 3000 million years before the first multi-celled creature appeared. Living things then evolved into separate Kingdoms, but because it happened so long ago, there are plenty of arguments over how many Kingdoms there are. In terms of the number of known species, the Kingdoms of the animals and of the plants dominate. All animals ultimately depend on photosynthesis by plants to produce organic molecules from sunlight, carbon dioxide, and water, with oxygen as a by-product; and animals recycle these materials back into carbon dioxide and water.

How much of the biomass produced by photosynthesis is captured by animals? This is a big question but we actually know the answer: in aquatic ecosystems more than 80% of the biomass from photosynthesis is eaten by animals, but much less, between 10 and 20%, of the products of terrestrial photosynthesis is eaten by herbivores.

If we look at all the animal species that we know about on Earth, more than 80% of them are either herbivores feeding on plants, or the natural enemies (mainly parasitoids) of herbivores. Thus we could say that the plant–herbivore–parasitoid interaction is the most important interaction on the planet.

Eating plants is not easy. Only a few mammals have managed it (ungulates, some primates), and hardly any birds feed on mature leaves—they usually select the protein- and mineral-rich fruits and seeds. Most herbivores are insects, but only 9 of the 26 Orders of insect feed on plants. Most herbivore species are beetles, but only one-third of beetles can feed on plants. Thus the ability to feed on plants is taxonomically restricted—it is hard to evolve to be herbivorous. Why is this?

Plant material is really tough. A herbivore has to fragment the plant tissues and break open the cell walls. To do this it needs specialized teeth, or their alternatives. Birds do not have teeth, but instead they have very powerful muscular crops and gizzards that rasp the husks from seeds. Many birds swallow stones and store them in the gizzard to help in this rasping process. A group of mammals—the ruminants—chews plant material twice: after the first chew, the material goes to the rumen, and then is regurgitated and chewed again.

Even when well chewed, plant material is extremely difficult to digest. Vertebrate herbivores do not have the enzymes to manage it, and digestive success depends on having microbes to do the job for them. Microbes need time to maximize the digestion. These requirements have led to the evolution of a range of organs that delay passage of the chewed material through the gut:

- the rumen of ruminants;
- foregut sacs in primates;
- a sacculated colon in horses;
- the caecum in many mammals (= the appendix in humans).

Even so, much of the material cannot be digested. Thus the success of digestion depends on having a large volume of microbe-laden fluid, a long time for the passage of ingested material in the gut, and the ability to eat a lot to compensate for the large proportion of totally indigestible material.

Insect herbivores differ from vertebrate herbivores in that they are mostly specialized to one or a few plant species. This means they can have a reduced set of gut microbial species, and more specialized enzymes tailored to the plants they feed on: hence they do not need long digestion times.

Plant defences

The consumption of plants by herbivores generates selection to reduce the impact of herbivory, and plants have evolved a whole range of defences against herbivores: mechanical, physiological, chemical, and phenological. The basic message is that not all green food is palatable. As Terry Pratchett put it: 'Behind every edible plant are numerous unsung cavemen who proved the hard way that others were poisonous'.

There are all sorts of mechanical defences present in various plants to try to deter herbivores. Against mammals there are spines, prickles, and thorns, and on the micro-scale, against insects there are hooked, sharp, or glandular hairs. Many small insects are stuck down by proteinaceous glues exuding from the ends of glandular hairs—once prevented from moving, they are then killed by the desiccating effects of the Sun.

If mechanical defences can be overcome, then a major problem for both mammal and insect herbivores is the structural properties of plants: the sheer toughness of leaves, and the indigestibility of tree bark. Have you ever wondered why plant cells walls are so thick and stiff? Between 80 and 90% of their mass consists of celluloses and hemicelluloses, which are extremely difficult to digest. There are some microbes that can manage it, but the distribution of cellulase enzymes is very patchy and uncommon. Plant structural compounds act as *digestibility reducers.*

Plants also contain a whole range of compounds that do a similar job: lignins, tannins, and the silica of grasses. Lignins form the 'woody' parts of woody plants: they are almost totally indigestible, and in addition they inhibit the herbivore digestive enzymes and puncture the guts of herbivores. The large molecules of tannins create the astringent taste of tea, and also inhibit or denature digestive enzymes by binding them together. Silica is basically glass, and is used as a structural molecule by grasses—it does not contain anything digestible, and unsurprisingly causes dramatic tooth wear in herbivores. These are all dosage-dependent *quantitative* defences—the more a plant has of them, the better defended it is. Another great advantage is that such defences are effective against *all* herbivores.

If digestibility is an issue, nutritionally plant material is not very good for animals either. The basic problem is that on average plants have only one tenth of the nitrogen of animal tissue. Nitrogen and phosphorus are often the limiting factors in animal nutrition, because nitrogen in particular is required to build proteins, and proteins are essential for growth. Thus, especially for juvenile animals, feeding on plants risks not obtaining enough nitrogen. Even if their digestive system is 50% efficient (and most are less than that), then an animal needs to eat twenty times its body weight in plant material just to get enough nitrogen. This does not even take into account the fact that the nitrogen in plants is often bound up in complex molecules that are especially hard to digest. All these plant features form a kind of biochemical defence against being eaten by herbivores.

A final major problem for herbivores is that many plants contain toxins that are specifically targeted *at them.* These act in very low quantities, and so are *qualitative* defences—if a plant has the right toxin for the right herbivore, then quantity doesn't matter. A general term for such toxins is 'secondary compounds'—a very strange name derived from the days when no-one knew what role they played in plant tissues. They were clearly not part of plant metabolism, and there were even suggestions that they were merely accidental by-products that had no function at all. Now we understand that they certainly do have a function—that of defence against particular herbivores. They are often concentrated in the reproductive parts of plants—the flowers, fruits, and seeds—precisely because these are the most important bits that need defending—they represent the fitness of the plant.

Plant toxins are hugely diverse—there are more than 14 000 kinds of terpene alone. We can group plant toxins into a number of major classes that include alkaloids, cyanogenic glycosides, glucosinolates, terpenoids and terpenes, and phenolics. The alkaloids include many familiar compounds, such as strychnine, nicotine, caffeine, morphine, cocaine, and quinine. If you cannot start the morning without a cup of coffee, then think about

Figure 4.2 The effect of caffeine on an orb-weaving spider.

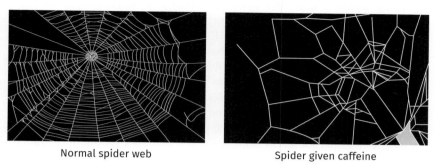

Normal spider web   Spider given caffeine

Redrawn from data from NASA/WikimediaCommons/Public Domain

what the caffeine is doing to you. Like all the others, caffeine is a poison directed at insect herbivores—it targets DNA and RNA synthesis. Some interesting experiments involved feeding small doses of caffeine to an orb-weaving spider—as Figure 4.2 shows, the effect on web-building was extraordinary. Many plant toxins are useful to us in all sorts of contexts, such as medicine (e.g. morphine), food (e.g. capsaicin in peppers), or agriculture (e.g. the pyrethroids used as insecticides).

Human parents are often worried about why their children refuse to eat vegetables. Commercial vegetables have usually been bred to have lower concentrations of poisons than their wild ancestors, but often they are still there. For example, many people dislike the piquant taste of brussel sprouts—which derives from the glucosinolate poisons they contain. We know from detailed studies that dislike and refusal of vegetables occurs in waves during childhood, with peaks corresponding to vulnerable stages of brain growth. Thus it makes perfect sense for a child to refuse to eat something that might damage its brain: perhaps it's the parents who are wrong!

We can think of plants as having two main strategies for deterring herbivores, depending on how large and long-lived the plant is. Large, long-lived plants such as trees will almost certainly be found by herbivores of all kinds, simply because the leaves are out there in the environment for a large proportion of each of many years. Because they cannot avoid being found, they need the generalized protection that comes from quantitative defences. Thus trees and grasses invest heavily in the quantitative defences of lignins, tannins, and (for grasses) silica. New buds and leaves have low levels because they need cell division and growth to expand in spring, but as soon as the leaf has expanded, the plant pours in tannins to toughen it up and defend it against being eaten. Thus the tannin content of a new oak leaf (*Quercus robur*) increases from less than 1% up to 6%, and simultaneously the protein content decreases from 40% when the manufacture of leaf material is in full swing, down to 15% when the leaf is mature. Even 1% tannin will stunt the growth of young caterpillars. Herbivores must hatch at exactly the right time in spring, just at bud burst. Too early, and there will be nothing to eat; too late, and the leaves will already be too tough for their delicate mandibles to cut and chew.

Other plants rely on not being found by generalist herbivores, by only being in leaf or flower for a short period of time, and 'hiding' in amongst other plants. They only need to worry about the specialized herbivores that are looking specifically for them. If found, then they only need a toxin tailor-made for that particular herbivore. Small herbs are typical of the plants that use this kind of strategy. They don't invest in quantitative defences, but instead have small amounts of toxins to deal with just a few specialized herbivores, which are usually insects. A short-lived grassland herb such as wild mustard (*Brassica napus*) contains no tannins and no lignin, but instead has low levels (<1%) of a set of toxins, the glucosinolates, and especially one called sinigrin. At levels of 0.001%, sinigrin stunts the growth of caterpillars of the Black Swallowtail butterfly (*Papilio polyxenes*), leading to abnormally small adults that lay few eggs: levels of 0.1% kill the caterpillars.

Herbivore adaptations

So what sort of herbivores can feed on a defended plant, if they all contain toxins? The answer is that many can. Wild mustard is a *Brassica*, and one of the commonest herbivores on brassicas is the 'cabbage white' butterfly—really a mixture of two species, the Large White (*Pieris brassicae*) and the Small White (*Pieris rapae*). As the name betrays, these species are brassica specialists—they actually use glucosinolates to find their host-plants! Their caterpillars grow perfectly well on food laced with sinigrin. What is going on?

Having poisons in their food puts selection pressure on herbivores to evolve mechanisms to deal with the poisons. Both vertebrate and invertebrate herbivores have developed a set of enzymes—the mixed-function oxidases—to deal with poisons. They occur in the liver of vertebrates and the guts of insects. Within the mixed-function oxidases are a related suite of enzymes called the cytochrome P450s, whose job it is to degrade dietary toxins. The fruitfly *Drosophila* has about eighty P450 genes producing these enzymes. Because it is their job to deal with novel chemicals that the insects ingest, it is not surprising that the P450s are also associated with the development of resistance to insecticides.

In addition to specific enzymes, resistance to dietary toxins often involves microbes in the gut, because microbes can often detoxify poisons more efficiently than animals can. Vertebrates in particular depend on microbes, but they occur in the guts of insects as well.

There is a *cyclical* process happening here. A plant develops a new toxin in response to herbivore-induced selection pressure. This puts the herbivore under selection pressure to detoxify it. If it manages to evolve this, then the pressure is back on the plant to develop something else new. This is the coevolutionary arms race when it involves one host-plant and one herbivore. When the whole community is involved in responding reciprocally, then it is known as the Red Queen hypothesis, from Lewis Carroll's *Alice in Wonderland* (Figure 4.3):

> "….in our country," said Alice, still panting a little, "you'd generally get to somewhere else if you ran very fast for a long time, as we've been doing." […] "A slow sort of country!", said the Queen. "Now

here, you see, it takes all the running you can do, to keep in the same place. If you wanted to get somewhere else, you must run at least twice as fast as that."

Figure 4.3 The Red Queen with Alice, in Tenniel's illustration of *Alice in Wonderland*.

Case study 4.1
Parsnips v parsnip webworm

A beautiful example of a coevolutionary arms race is the interaction of the Parsnip and its specialist herbivore, the parsnip webworm, a pair of species imported from Europe and studied by May Berenbaum in the USA. Parsnip (*Pastinaca sativa*) belongs to the Apiaceae, whose characteristic poisons are the furanocoumarins. These are concentrated in the reproductive parts, with especially large amounts in the fruits and seeds. The parsnip webworm (*Depressaria pastinacella*) is a moth whose caterpillars feed gregariously in the flower clusters of parsnips, actually preferring to eat the flowers and seeds.

There is plenty of genetic variation in toxin levels among individual plants, and also in P450 enzyme activity levels among individual caterpillars. The caterpillars destroy the flowers and seeds of the plant, which are its reproductive output, and so damage the evolutionary fitness of the plant. However if the plant escapes being found by egg-laying female webworm moths, then it is actually better *not* to have toxins because the metabolic process of making them diverts resources from flower and seed production. Thus in the absence of webworms there is selection against toxin production, whilst in the presence of webworms there is selection for toxin production.

CS4.1 Figure A The parsnip and parsnip webworm system.

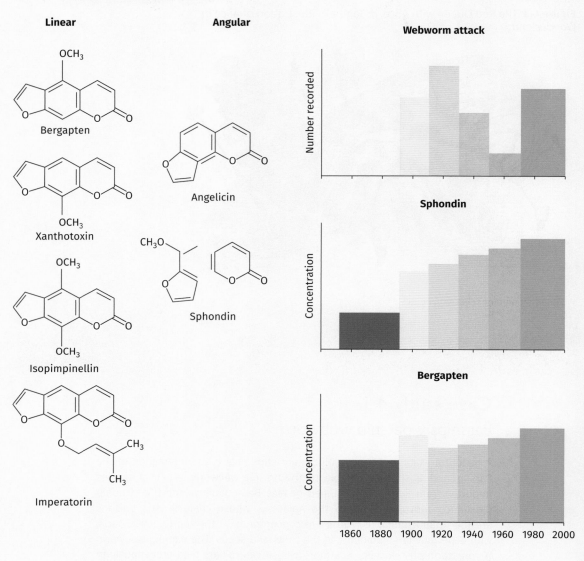

CS4.1 Figure A Continued

(ii) Bildagentur Zoonar GmbH/Shutterstock.com, (iii) Naturepix/Alamy Stock Photo

When parsnips were first introduced to the USA from Europe for food in 1609, the webworm did not come with it. For 260 years, parsnips grew in North America with no specialized herbivores at all. The first webworm was recorded in 1869, then it spread rapidly to infest most of the plant populations. In a clever experiment, Berenbaum checked the toxin levels and webworm infestation levels of dated herbarium specimens of parsnip plants. You can see her findings in Figure A. Amazingly, most of the toxins had disappeared from American parsnips by the early nineteenth century, but once webworms came on the scene, levels rapidly increased so that by the end of the twentieth century they had reached levels similar to those in Europe.

❓ Pause for thought

What are the dangers of moving organisms around the world, as humans do inadvertently all the time?

Figure 4.4 Evolutionary trees of milkweed plants (*Asclepias* spp.) and their specialist herbivores, the milkweed beetles (*Tetraopes* spp.).

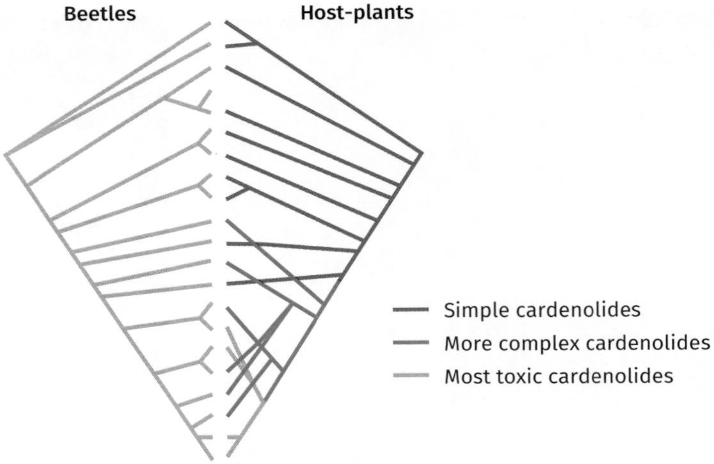

Reproduced with permission from Farrell, B. D., Mitter, C. The timing of insect/plant diversification: might *Tetraopes* (Coleoptera: Cerambycidae) and *Asclepias* (Asclepiadaceae) have co-evolved? *Biological Journal of the Linnean Society* 63: 553-77. Copyright © 2008, Oxford University Press.

The result of such coevolution should be the generation of new species and escalation of the arms race, producing ever more toxic poisons through evolutionary time. A nice example of this concerns species of milkweed (*Asclepias* spp.) and their specialist beetle herbivores (of the genus *Tetraopes*) (Figure 4.4). Milkweed plants are so-called because of the milky juice that exudes when they are damaged, laden with alkaloids called cardenolides. You can see from the pattern of the phylogenies that speciation on each side of the arms race is linked. Furthermore, the evolution of the plant has involved the invention of increasingly more complex and more toxic cardenolides, just as predicted by the theory of arms races.

Predators and trophic cascades

The cheetah (*Acinonyx jubatus*) is a large and very charismatic carnivore that faces particularly acute difficulties in living alongside humans. It now lives in only 9% of its former range with only about 7100 individuals left. With an estimated rate of decline of 10–13% per year, the outlook for the cheetah is gloomy.

Does it matter if the cheetah is driven to extinction? Top predators like the cheetah have similar features in being large, reproducing slowly, and living at low densities, all of which make them more vulnerable than usual to extinction. Thus the more general question is: does it matter ecologically if top predators are lost? Of course, it matters a lot to *us* if they disappear because all of them are species that we value highly, and many of us will spend a lot of money to go to see them. But does it matter to the ecology of the area?

The answer is *yes*—it matters a lot. Although herbivore–plant relationships are numerically the most important in nature, predator–prey relationships are also very important ecologically, and can alter landscapes completely. The best way to illustrate this is by describing in some detail an example from the western United States.

Wolves and Yellowstone National Park

Established in 1872, Yellowstone National Park was the first of its kind in the world, kick-starting the conservation movement. The creation of the park also resulted in the expulsion and exclusion of its indigenous people, something that was to become depressingly normal all over the world during the next 120 years.

From the middle of the twentieth century, biologists and park managers in Yellowstone noticed changes in the woodlands along streams and rivers, especially in valleys where Red deer (*Cervus elaphus*, called 'elk' in North America) congregate in the winter. Poplar (*Populus* spp., called 'cottonwood' in North America) and willow (*Salix* spp.) trees were disappearing and the landscape was opening up to become grassland. What on earth was happening? There seemed to be no recruitment of new young trees to these populations.

Robert Beschta and William Ripple were called in to study the problem. From the age distribution of poplar trees in Figure 4.5(a) they deduced that there had been a tree regeneration problem in these areas: there are no surviving poplar trees after about 1960, but the problem started at about 1900–1920 when recruitment started fading away. There were plenty of seedlings—up to 70 000 per hectare. Thus seeds were reaching the right places, and were germinating, but they were not surviving to produce adult trees.

Alarmed by the loss of trees, park managers established fenced exclosures in the period between 1930 and 1950, and over the years, strong differences in the vegetation inside and outside these exclosures developed. Outside the exclosure plots, there were no trees at all, whilst inside them a dense thicket of trees grew up. Clearly, pressure from grazing and browsing had something to do with the lack of tree regeneration. After 1998, this pressure suddenly relaxed, and the poplar and willow trees started growing again outside the exclosures.

Red deer are the main grazer/browsers in Yellowstone, so why did their impact increase in the early twentieth century, then decrease again at its end? Figure 4.5(b) shows that the changes correspond with the loss of wolves from the park in the 1920s and their return in 1996. After the establishment of the park there was a period when 'predator con-trol' was in vogue, and gradually all the wolves were killed, the last ones being eliminated in the mid-1920s. We know that Red deer, especially the calves, are one of the favourite prey of wolves. We might predict that when wolves disappear, the populations of Red deer will soar and hence the grazing pressure will increase too; and when the wolves come back again, Red deer populations will drop and so will the grazing pressure. However, when you look at the actual data in Figure 4.5(b), that is not what happened. Deer populations actually drop consistently during the period without wolves.

Figure 4.5 The disappearing trees of Yellowstone Park. (a) The age profile of poplar trees in Yellowstone Park in groves in the river floodplains; (b) reconstructed populations of wolves and elk.

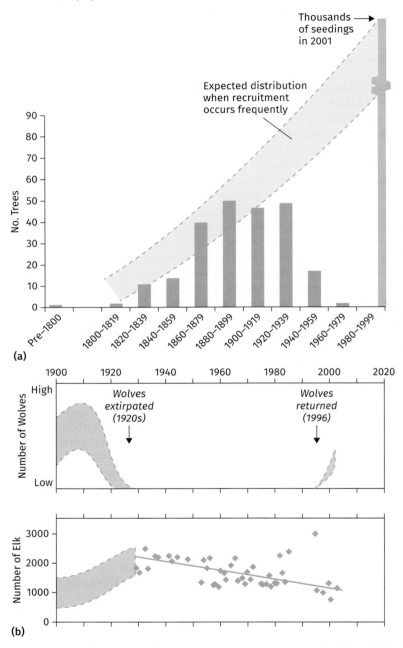

(a) Reproduced with permission from Robert L. Beschta. Cottonwoods, Elk, and Wolves in the Lamar Valley of Yellowstone National Park. *Ecological Applications*. Oct 1, 2003. Copyright © 2003, John Wiley and Sons. (b) Reproduced with permission from William J. Ripple, Robert L. Beschta. Wolves, elk, willows, and trophic cascades in the upper Gallatin Range of Southwestern Montana, USA. *Forest Ecology and Management* 200: 167. Copyright © 2004, Elsevier. DOI: 10.1016/j.foreco.2004.06.017

So the presence of wolves affects grazing pressure, but not by dramatic changes in deer numbers: how might this work? A clue lies in the spatial variability in vegetation recovery. In the woodlands alongside the river, the recovery of the trees is obvious. But in the open uplands, there are no signs of vegetation recovery from grazing—but then these areas were not much affected in the first place. The most logical conclusion is that Red deer are scared of wolves, so they feel unsafe in places where wolves can creep up on them unnoticed. If wolves are present, then deer avoid the riparian woodlands, remaining in places with good visibility where they can see the wolves coming.

This is the concept of the 'landscape of fear' created by top predators: some are capable of reducing prey populations directly, but they can certainly also have a strong effect indirectly by creating fear in their prey and hence a change in behaviour. This latter effect was sufficient to change the appearance of the Yellowstone landscape completely, switching riverside woodlands into open grassland. The loss of wolves was not restricted to Yellowstone, but occurred throughout western North America, and the change in vegetation was replicated everywhere.

The impact of top predators on lower levels in the trophic pyramid is known as a trophic cascade, because the effects often (but not always) cascade down the trophic levels. They can take different forms in different communities, but the negative effect of the top predator on the level below acts to the benefit of the next level down.

The famous lynx–hare cycles of Canada described in Chapter 1 form another example. The top predator is the lynx, whose favourite prey is the snowshoe hare: lynx predation is the main cause of their mortality. The hares feed mainly on shoots of willow and birch. The fundamental cause of the cycles is once again the 'landscape of fear'. High densities of feeding hares reduce the availability of twigs, forcing hares to spend more time eating lower-quality food, exposing them to a greater risk of predation. This makes them frightened and stressed, which leads to fewer offspring who themselves have fewer offspring: this maternal effect lasts for two generations. Thus the hare cycles are driven by the mortality impact of predation, and the indirect impact of fear on reproduction. The lynx cycles are driven by variation in their food supply (the hares), low levels causing delayed reproduction and movement—which leads to the 'spatial ripples' across the Canadian landscape.

Indirect interactions

Thirty years ago, Andrew Sih and his colleagues reviewed experimental studies of the impact of predators, i.e. those that experimentally manipulated the presence or the density of predators and looked at impacts on their prey. There turned out to be plenty of strong evidence for the importance of predation in nature. However, in more than half of cases, although strong, the impacts were also 'unexpected', i.e. prey densities did change, but not in the predicted direction. For example, when a predator was removed, prey density *decreased*. Why was this?

The main reason is that organisms do not simply have one-to-one feeding relationships, but instead are embedded within a community, a **network** of interactions. In reality, all organisms are embedded in such networks, and therefore there is potential for a huge variety of indirect effects. We know now that such indirect effects are extremely common in natural communities, with both powerful and subtle effects.

A good example is shown by the impact of a couple of European invasive species on a native food web in the Patagonian forests of South America (Figure 4.6a). The food web is based around a dominant

Figure 4.6 (a) Food web based on the shrub *Aristotelia chilensis* in Patagonia (Chile and Argentina); (b) indirect effect of exotic ungulates on hummingbird pollinators via browsing on *A. chilensis* and hence reducing mistletoe numbers; (c) direct effect of wasps on the number of fruits (hence seeds) dispersed by birds and marsupials.

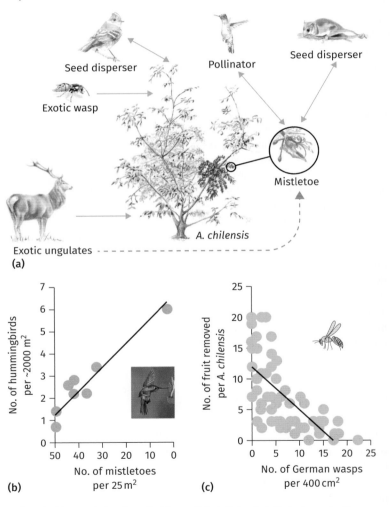

Reproduced with permission from Rodriguez-Cabal, M. A., et al. Node-by-node disassembly of a mutualistic interaction web driven by species introductions. *PNAS* October 8, 2013. 110 (41) 16503-16507; [https://doi.org/10.1073/pnas.1300131110]. Drawing by Ezequiel Rodriguez-Cabal. Hummingbird image in (b) Patagonian Stock AE/Shutterstock.com.

understorey shrub *Aristotelia chilensis* and a mistletoe *Tristerix corymbosus* that grows on it. Mistletoes are plants that parasitize and damage other plants (usually trees or shrubs) by growing on them and stealing water and nutrients. This mistletoe belongs to a group whose flowers are pollinated by hummingbirds, in this case by the Green-backed Firecrown *Sephanoides sephaniodes*, which pollinates one fifth of all the woody plants in Patagonian forests. The hummingbirds depend on the nectar-rich mistletoe flowers over the winter, and are almost the only pollinators—this is therefore a relationship of mutual dependence.

South America is home to about a quarter of the world's marsupials (which are of course mostly Australian, but South America was probably their original home), and an endemic species, the Colocolo opossum *Dromiciops gliroides*, feeds mainly on the mistletoe fruits and is the only known disperser of the seeds. After passing through the gut of the Colocolo, the seeds are deposited in the faeces on the branches of the shrub, where they germinate. A bird, the White-crested Elaenia *Elaenia albiceps*, belonging to the family of tyrant flycatchers, disperses the seeds of most of the woody plants in the forest, including those of *Aristotelia* shrubs. Bird-dispersed seeds are three times better at germinating than seeds that drop from the plant.

European cows and Red deer *Cervus elaphus* were introduced into this environment more than one hundred years ago, and are now widespread in more than half of all the forests. They prefer feeding on *Aristotelia*, and their grazing, browsing, and trampling reduce its density by 16 times, almost to zero, and with the shrub goes the mistletoe—its density is 83-times lower in invaded areas. The entire plant community is affected—total plant cover is 35-times lower, and habitat complexity is 20-times lower in invaded sites.

These changes affect the populations of the marsupial opossum, eliminating it completely from the invaded sites. Since mistletoe seeds need to pass through its gut, mistletoes are unable to recruit new plants into the population. The abundance of hummingbirds is directly related to the density of mistletoes, so if the mistletoes disappear, so do the hummingbirds (see Figure 4.6b).

If that is not bad enough, the habitat has taken another blow in the arrival and very rapid spread of German wasps *Vespula germanica* from Europe in the 1980s. The wasps remove and feed on *Aristotelia* fruits, which can easily be reduced threefold in availability to the birds and marsupials (see Figure 4.6c).

Both direct and indirect interactions can lead to cascading linked extinctions and the collapse of native foodwebs, and in this case for some elements the indirect effects are stronger than the direct effects. We interfere with these webs at our peril!

The complicated case study shows that the community context really matters. For conservation, therefore, trying to conserve the populations of a single species usually means taking measures that sustain the entire community. It is hard to know what subtle interactions occur within communities, and hence who depends on whom for survival. Heavy-handed interventions by humans can easily ruin such dependent relationships.

Case Study 4.2
Aphids, hoverflies, and poison

This case study concerns a food web in the South Sinai mountains based on the herbivores feeding on a single species of plant, the Sinai milkweed *Asclepias sinaica*. Because it belongs to the milkweed family, it has physical defences in the form of canals filled with latex under pressure, so that when damaged, sticky latex exudes and then hardens, trapping small herbivores. There are also defensive poisons—cardiac glycosides called cardenolides—in the plant tissues, especially the seeds and fruits. These stop the heart, and hence are toxic to most vertebrates. Cardenolides are toxic also to most insects, and hence are an effective deterrent to insect herbivores as well. However, individual plants vary a lot in toxin content, so it is possible to choose less toxic plants on which to feed.

There are four main herbivores specializing on Sinai milkweed and its relatives, all of them insects: a weevil *Paramecops sinaitus*, a bug *Spilostethus pandurus*, an aphid *Aphis nerii*, and a grasshopper *Poekilocerus bufonius* (Figure A). All four store the plant poisons in their own bodies for use in their own defence. Vertebrate herbivores such as ibex, camels, and goats steer well clear of these plants.

This case study focuses on the feeding relationships of the aphid, and in particular the feeding behaviour of its hoverfly predator. Simberloff's challenge was to try to explain features of one level of organization from lower levels—the reductionist approach. So we ask the question: do we need the 'community' level in order to understand the feeding behaviour of the aphid-feeding hoverfly?

About 40% of all the world's 6000-odd species of hoverfly have predatory larvae, mostly feeding on aphids. Hoverflies are true flies, and hence have blind, legless larvae. The predatory larvae are usually brightly coloured, often with beautiful cryptic patterns of yellow, red, green, and black. They like moist conditions, so we would not expect to find many in the driest country in the world. In fact, there are only 55 recorded species of hoverfly in Egypt, but 39 of them have aphid-feeding larvae. By comparison, the UK has about 265 hoverfly spp. Egypt is about four times bigger than the UK, yet the UK has about five times as many species of hoverfly.

One hoverfly feeding on milkweed aphids in Sinai is called *Eupeodes corollae*. It is a very common species across the whole of Europe including the UK. The larvae have been found feeding in the colonies of a wide range of aphid species. Because they don't have legs, the larvae cannot move between plants and hence are stuck with the plant on which the egg was laid. This means the selectivity of the gravid (pregnant) female is crucial—she must choose wisely otherwise her offspring will die. So our question is: do we need the community context in order to understand what influences her choice?

Larvae have to be able to feed on the aphids, so females must choose a colony of palatable aphids. Aphids defend themselves in a variety of ways against attack by predators, including running away, secreting sticky droplets, rapidly dropping from the plant or having chemical toxins of a wide variety of types. *Aphis nerii* is bright yellow, a sure indication that it doesn't taste very nice. Its distribution on plants is not related to the levels of milkweed poisons: it is not poisoned, but stores the toxin for its own use, so its level of defence matches that of the plant. However, even the delicate neonate (just hatched)

CS4.2 Figure A Food web based on Sinai milkweed.

ladybird *Adalia*

hoverfly larva
Eupeodes

parasitoid
Diplazon

aphid
parasitoid

Milkweed weevil
Paramecops

Oleander aphid
Aphis nerii

Crematogaster ant
tending *Aphis nerii*

Seed bug
Spilostethus

Sinai milkweed
Asclepias sinaica

Milkweed
grasshopper
Poekilocerus

(i) Image © Fred Manata, by kind permission, (ii) Image © Stine Simensen, by kind permission, (iii) Image © Fred Manata, by kind permission, (iv) David Cappaert, Bugwood.org, (v)-(vi) Image © Fred Manata, by kind permission, (vii) Image © Jen Johnson, by kind permission, (viii) Doug Wechsler/Animals Animals/agefotostock, (x) Henri Koskinen/Shutterstock.com, (xi) © Steve Hopkin/ardea.com

hoverfly larva appears to be able to feed on *Aphis nerii* quite happily if the sequestered toxins are not too concentrated, so a female laying an egg in such an aphid colony is making a reasonable decision. She does not lay in the nearby colonies of *Aphis verbasci* on *Verbascum* because although its toxin will not kill her larvae, it is not digested well and slows larval development, resulting in smaller adults that will not lay many eggs. So although *Eupeodes corollae* larvae can survive on verbascum aphids, their overall fitness is reduced.

Most of the mortality of hoverfly larvae comes from being parasitized by wasps: there is an entire subfamily of 350 wasp species called the

Diplazontinae that are completely specialized to parasitize hoverfly larvae. The female wasp searches in likely habitats for the aphid colonies in which her prey are found. When she comes across a hoverfly larva, she stabs it with her ovipositor and lays her egg inside in the body cavity (or sometimes just behind the brain). Such insect parasites are called 'parasitoids' because their larva eats out the host completely before pupating and emerging—thus the host always dies unless its defence mechanisms manage to kill the parasitoid egg before it hatches. Some egg-laying decisions of a female hoverfly are based on whether her larvae are likely to die from parasitoid attack. Because female diplazontines search for particular plants and particular aphids, this likelihood varies among different plant and aphid species.

Hoverflies are not the only insects that feed on aphids, so there is competition for aphid prey to take into account. The main competitor on an individual milkweed plant is sometimes another hoverfly larva (many are cannibals!), but usually is a ladybird. Most aphid-feeding insects will try to gobble up another if encountered, so who eats whom depends largely on size and power. The most vulnerable stage of the hoverfly is the egg or the young neonate larva, so there is no point in laying an egg in an aphid colony that already has a ladybird or another hoverfly in it. Gravid females check carefully before laying their eggs: a wise hoverfly mother therefore chooses to lay her egg where the risk of cannibalism or predation is lowest.

The last major factor a gravid female hoverfly needs to take into account is the occurrence of ants, some of the most ferocious predators of the micro-world. Ants exploit aphid feeding behaviour, while aphids exploit ants to defend them. Aphids feed from the sugar-carrying vessels (the phloem) of plants, but what they want is the low concentrations of nitrogen, not the super-abundant sugar. Thus they excrete large quantities of sugar solution ('honeydew') in order to extract the nitrogen from the flow. Ants 'farm' particular species of aphids by keeping off their predators and parasitoids, in return for producing honeydew on demand, i.e. when stroked by ant antennae. Some aphids are nearly always attended by ants, whilst others never are.

The presence of ants makes a big difference: they will fend off gravid female hoverflies, preventing them from laying their egg; and they search out and kill eggs and young hoverfly larvae. Not surprisingly, this creates strong selection in hoverflies for countermeasures, and there is an entire genus of hoverflies—*Paragus*—that are specialized for evading ant defences. They have very persistent oviposition behaviour, lay extremely hard-shelled eggs and their larvae have long spines. This combination of traits mean the larvae can overcome the defences of some ants long enough to take on the odours of the aphid colony, and from that point onwards they are safe—they are not recognized as foreign by the ants. *Eupeodes corollae*, however, has none of these adaptations, and hence cannot cope with ant-tended colonies of milkweed aphids.

Thus a female hoverfly *Eupeodes corollae* has a lot to weigh up when laying her egg. She has to choose the right aphid on the right plant, avoid ants and competing aphid-feeding insects, and select somewhere that minimizes the risk of being parasitized by a diplazontine. Only by seeing the world through the eyes of the female hoverfly and understanding the community context can we hope to understand the feeding behaviour of the insect.

 **Pause for thought**

Insects are mostly very specialized compared to mammals and birds. Do you think the ideas we have explored here apply to vertebrate food webs?

Chapter summary

- Most species interactions in ecology are between insects and plants.
- Selection for defence against herbivores has led plants to evolve toxins.
- Selection for the ability to use a toxic diet has led herbivores to evolve counter-measures, such as the P450 detoxification enzymes.
- This coevolutionary process is a major source of evolutionary change.
- Top predators can have disproportionate effects (a 'trophic cascade') on the species below them in the food web of their community.
- The eradication of wolves by humans in Yellowstone National Park had the unintended consequence of preventing tree regeneration.
- The effects of top predators can be direct (on prey populations) or indirect (e.g. on prey behaviour).
- All species are involved in interactions within communities.
- Many such interactions are indirect via intermediate species.
- Conservation needs to pay attention to the variety of ways in which such indirect interactions work.

Further reading

Levin SA (ed) *The Princeton guide to ecology*. Princeton UP. pp. 227–232, 247–252.

The Princeton guide consists of short (4–8 pages) chapters written by different authors outlining concepts and ideas in ecology and conservation. Some are rather detailed, but they expand and consolidate the ideas presented here.

Discussion questions

4.1   What are some of the implications of the ideas discussed in this chapter for conservation?

4.2   The UK has a major problem of tree regeneration caused by deer populations being too high. What solutions might there be?

5 WHAT PROCESSES CREATE ECOLOGICAL COMMUNITIES?

In 1999 a famous ecologist, Professor (now Sir) John Lawton, claimed that the main problem with ecological ideas and ecological studies was that they were overwhelmingly local in nature. The vast majority of studies involved a pond, a field, or a small part of a wood, hedgerow, or grassland. The implication was that we needed to expand in scale, to add to our understanding the large-scale processes that are undoubtedly important (Figure 5.1). Since then there has been a vast increase in thinking and studying at these large scales.

Figure 5.1   A single tree is an ecosystem, home to hundreds if not thousands of different species. But do the rules of this small ecosystem also apply to huge rainforests, mighty oceans, and seemingly endless deserts?

(a) Mike Richter/Shutterstock.com, (b) katatonia82/Shutterstock.com

In this chapter we describe the impacts of scale on ecology. We do this by contrasting what ecologists think are important at the local scale with processes at the regional scale. They are connected, of course, and the form of these connections is also important.

Local processes

What determines whether a species is regularly present at any particular spot on Earth? Obviously for a species to be present with any degree of regularity (e.g. permanently), it has to be available in the regional pool of species. It has to be able to reach the site, and to survive, grow, and reproduce there. This means that the environmental tolerances of the organism need to be considered. If we take temperature, for instance, there are several different possibilities:

- in many places it can be too cold or too hot for individuals to survive, hence they do not exist in that environment;
- the temperature may allow individuals to survive temporarily, but be too hot or too cold for them to grow;
- the temperature may enable individuals to do more than survive—they may be able to grow, but the conditions are too extreme for reproduction;
- the temperature may enable organisms not only to grow but also to reproduce—but even that doesn't guarantee a species will survive for long;
- the temperature may allow *enough* reproduction to take place for population maintenance, to prevent the population dying out.

Thus there will be a 'best' range of temperatures for crucial functions such as reproduction, somewhere in between the tolerance limits, that allows a species to maintain its population. Somewhere there will be a temperature that maximizes performance (the combination of survival and reproduction)—the *optimal* temperature conditions (Figure 5.2).

What we are describing is the temperature niche of the organism. This kind of definition of the niche may not be the one you are used to (the 'role' of an organism)—we will explore the relationship between the two shortly. One of its great advantages is that it is easily measured. Furthermore, each organism is affected by a whole range of environmental factors which help to define its ecological niche. In principle we can define this curve (Figure 5.2) for any environmental gradient—temperature, water availability, soil pH, etc.

Figure 5.2 The ecological niche of an organism is where it cannot only survive, grow, and reproduce, but also maintain a population.

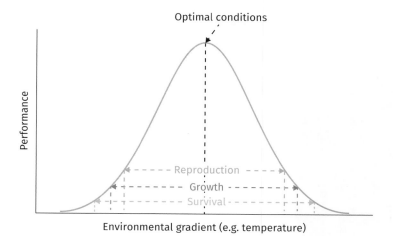

A concrete example (Figure 5.3) will help. In ponds there are many small invertebrates, among which are crustaceans called water fleas (*Daphnia*, see Figure 5.3a)—these row themselves jerkily around in the water using their appendages, and feed by filtering unicellular algae, protozoa, and bacteria from the water. *Daphnia* numbers are crucial for the many other species that feed on them. Ecologists thought that pH and calcium-ion concentrations were the main environmental factors influencing the distribution of *Daphnia* in the ponds they observed, and set up lab experiments to test their ideas.

It turned out that there was a remarkable match between the lab experiments (Figure 5.3b) and the field data (Figure 5.3c). The experiments measured the 'performance' of *Daphnia* in different combinations of the values of two environmental gradients (pH and calcium concentration). Performance was measured as the population growth rate per day, and can be positive or negative—long-term population persistence requires the population growth rate to be above zero.

The occurrence of *Daphnia* populations in natural ponds turned out to be determined largely by the values of these environmental parameters. This confirmed the idea that every species has a niche, which is the combination of environmental parameters in which its populations can maintain positive growth. The experiment identified the most important parameters for *Daphnia*.

Abiotic or biotic factors?

The concept of the niche illustrated above is abiotic, because it is all about tolerances to environmental variables such as temperature, pH, and cation concentration. However, the idea of the niche has undergone some important changes during its one-hundred year history. In large part this was due to an

Figure 5.3 The ecological niche of the water flea *Daphnia* along the environmental gradients of acidity of the water (pH, a logarithmic scale) and the degree of hardness of the water (the concentration of the calcium ion, also on a logarithmic scale): (a) laboratory experiments at different experimentally created combinations of pH and calcium-ion concentration, smoothed to give a contour map describing the population growth rate (PGR) per day (the shading); (b) samples from natural ponds where their pH and calcium-ion concentrations were measured, and that do (solid) or do not (open symbols) contain *Daphnia*. (Circles and triangles represent two different datasets.) The dashed line is where the population growth rate is zero from the experiments, and the solid line where it is >0.2.

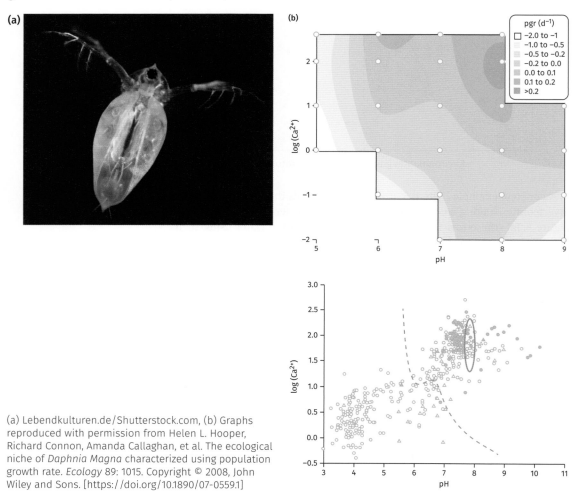

(a) Lebendkulturen.de/Shutterstock.com, (b) Graphs reproduced with permission from Helen L. Hooper, Richard Connon, Amanda Callaghan, et al. The ecological niche of *Daphnia Magna* characterized using population growth rate. *Ecology* 89: 1015. Copyright © 2008, John Wiley and Sons. [https://doi.org/10.1890/07-0559.1]

appreciation of the power of interactions between species, particularly competition. We now think of each dimension of the niche more as a resource required by an organism, rather than a tolerance to an abiotic variable such as temperature. A key point is that resources can be used up and become limiting to growth.

The original abiotic concept is now called the fundamental niche, whilst the actual niche occupied in the real world is called the realized niche. The two don't match because of the impact of competition with other species—thus the realized niche comes about because of the biotic rather than the abiotic environment. It is this concept that gave rise to the idea of the niche as the 'role' of an organism in its community.

Consider six species sharing environments that can be described along two niche dimensions, where each is a resource—Figure 5.4 illustrates this model.

The niche of each species in this idealized situation has an identical size and shape—following the *Daphnia* example, we draw only the contour of population growth rate within which the growth rate is positive, so population persistence is possible. However, because there is obvious *overlap* in resource requirements, this generates competition if the availabilities of the two resources are in short supply. When species compete for limited resources, then the better competitor will exclude the worse competitor from those resources—this is Gause's Principle, or the Principle of Competitive Exclusion. In the regions of overlap, therefore, one species will succeed in obtaining the resources whilst the others will be denied resources. If the red species, for instance, is a poor competitor, then the only resources available to it lie in the central part of the niche space that it alone occupies. The realized niche is then what is left over after the result of interactions with other species in the community.

To study the fundamental niche, therefore, you can study a single species in isolation, and measure performance along environmental gradients in order to understand the factors that limit its distribution and abundance. In contrast, to study the realized niche, you have to measure how a set of co-existing species share the available resources, and understand the competitive relationships between all the pairs of species. The community context is vital.

Many ecological studies concentrated on what was called 'resource partitioning by coexisting species' in order to understand the 'coexistence problem', i.e. how species avoid competitive exclusion. An overview of the results suggests there are three main niche axes where species differ: *habitat*, *food*, and *time*. An interpretation of this is that the main way in which coexisting species reduce competition is by occupying different habitats or

Figure 5.4 The realized niche. Each circle is the fundamental niche of a species. The six species overlap in their niche requirements, causing competition. The outcome of competition results in the realized niche—that proportion of the fundamental niche that a species actually occupies due to the effects of competitive interactions.

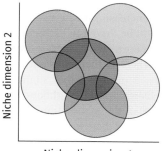

microhabitats; if they are in the same habitat then they coexist by feeding on different foods; and if they live in the same habitat and feed on the same food, then they coexist by some kind of temporal segregation—coming out at different times of the year, or different times of the day.

The role of competition

This theory of coexistence depends utterly on the importance of competition—this has to be the dominant force in communities in order for it to be the underlying process that causes the pattern of resource partitioning. How realistic is this? Ecologists have used experimental manipulations (removing or adding species to plots, or changing densities) to look at this issue—we have already seen one amazing example in Simberloff's mangrove experiments in Chapter 2. There have been hundreds of such studies, hence we can review them to assess whether competition really is as important as the theory claims.

In one review of 164 experimental studies of competition, more than 90% demonstrated significant impacts of manipulating the density of one species on other coexisting species. The studies involved 390 species, and more than 70% sometimes showed significant effects, and more than 50% *always* showed them in all the studied sites and times. Reviews such as these demonstrate that competition is definitely a potent force in most communities, affecting most species most of the time, and in most habitats where they occur.

If species need to avoid too much competition in order to maintain enough niche space to permit the population growth rate to be positive, then how many species can co-occur in a habitat? Surely there is a limit caused by the variety of the resources (i.e. length of the niche axis) relative to the size of the niche of each species? There is some elegant theory that deals with this issue (although as always it makes unrealistic assumptions to simplify the mathematics).

We assume that the number of species that can coexist will depend on how wide the niche of each species is, because we can fit in many more with narrow niches than we can those with wide niches. The theoretical conditions are shown in Figure 5.5. There is a set of identical niches differing only in their position along the niche axis, and a constant supply of resources at the same level along the axis. We then use mathematics to see what happens when we push the niches together, packing them more tightly. At some point the increasing niche overlap causes too much competition, and some species will be driven to extinction.

That point turns out to be when the distance apart is the same as the width of the niche (see Figure 5.5b). There is then substantial overlap between adjacent species along the axis—it turns out that slightly more than half of the resources (54%) can be shared with a neighbour before competitive exclusion kicks in. Even if we relax the assumption of constant resources and allow these to vary unpredictably, there is still a limiting similarity to the niches caused by competitive exclusion—with unpredictable resources, niches have to be less similar, more spaced, and with less overlap, but the limit is still there.

Figure 5.5 Niche packing and limiting similarity: (a) take a set of identical species, each with a niche of width w (the standard deviation of the normal curve) and separated by a distance d; using theory we now push the niches closer and see what happens in the population equations; (b) at some point the overlap is too great for coexistence, and the community will begin to lose species via competitive exclusion.

(a)

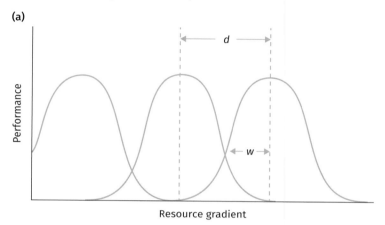

(b)

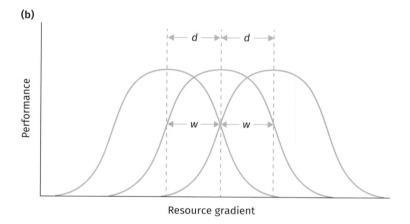

This is an interesting result, because it predicts that in a **saturated community** (i.e. one where no more species can fit in) of coexisting species dominated by competition, the species will be forced to space themselves regularly along a niche axis. Is there any evidence for this? Well, there certainly is some, although if competition really is a universally potent force there ought to be many more—an indication that other processes are often important too.

Figure 5.6 shows the results of one study of three rodent communities in the deserts of the western United States, where experiments show that rodents compete for seeds. The mean body mass of each of the four species in each community is plotted on a logarithmic axis. This is clever because

Figure 5.6 The sizes of rodents from three communities of the deserts of western USA (note the log scale).

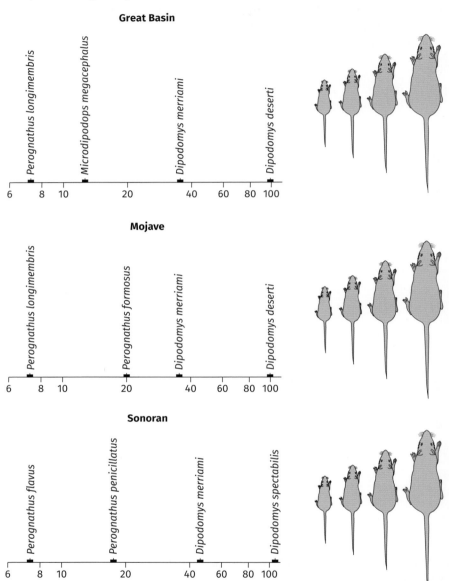

Adapted with permission from Bowers & Brown. Body Size and Coexistence in Desert Rodents: Chance or Community Structure? *Ecology*, Vol. 63, No. 2 (April 1982), pp. 391-40. Apr 1, 1982. Copyright © 1982, John Wiley and Sons [https://doi.org/10.2307/1938957]

of course a difference in log size is the same as a ratio (because division is represented by subtraction with log-transformed numbers). Thus what we can see here are more or less regularly spaced distances along the body-size axis, indicating the same ratio of body mass between adjacent species (the larger is about twice as heavy as the smaller). Indeed, statistically they are more similar than random expectation.

Since the cube root of two is 1.3, if body mass doubles then we would expect length ratios to be about 1.3:1. This ratio has been found—for example, in the lengths of the feeding apparatus of coexisting water boatmen. But it also occurs outside biology—in the lengths of the recorder family (Figure 5.7), in the sizes of children's bicycles, violins, and kitchen pans! You can see how the explanation might go—it is the kind of size difference that makes the function usefully different, in both inanimate objects and in competing species within communities.

The most impressive achievement resulting from the application of the theory of species packing also concerns desert communities, but this time of lizards. Since the two main axes of the niche of these animals are habitat and food, Eric Pianka decided to roam the world studying these aspects of desert lizards. The great thing about deserts is that the entire biological system depends on the availability of water, so we can use rainfall records as a proxy for the way in which resources vary from year to year. Pianka visited 28 desert sites in three continents (North America, South Africa, Australia), recording between 4 and 40 species per community. In general the North American sites had the fewest species (4–10) and the Australian sites had the most (18–40). Using census walks, he measured where each individual of each species was in the habitat. Over many such walks, he built up a picture of how each species used the habitat, and hence could measure the overlaps in habitat use between species. He also sampled individuals and looked at their stomach contents. This allowed him to build up another picture of what each species was feeding on, and hence measure the overlaps in food use.

Two Dutch researchers then took Pianka's data and added in long-term rainfall records to estimate the variability of the resources available to lizard communities in each site. Using the same theory of limiting similarity described above, they calculated what the relationship should be between average niche overlap and species richness if every lizard community was (a) saturated with species; and (b) competing for resources along the two main niche axes of habitat and food. Figure 5.8 shows the match between the real data and what the theory of limiting similarity predicts. As you can see, there is a clear match between theory and reality. This was an astonishing result. What it suggests is that the ideas of the niche, of competitive exclusion, and of limiting similarity accurately describe the limits to the number of species in local desert lizard communities all over the world.

Of course, this is all very simplistic and we cannot claim that it describes every community well, or even many of them. For example, the way resources are affected by consumption needs to be considered. As each individual consumes the resources it needs, the levels of those resources are reduced. What then matters is which species can survive on which level of each resource, and what the rates are at which the resources are renewed within a habitat.

Figure 5.7 The length ratios of adjacent recorders is about 1.3:1.

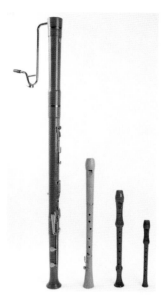

© chrisstockphoto/Alamy Stock Photo

Figure 5.8 The ecological niches of desert lizards such as the North American ornate tree lizard seen below. The mean niche overlap is the average of the overlap along habitat and food axes. The symbols denote the observed values in communities from three different continents. The solid line is the predicted relationship from niche theory.

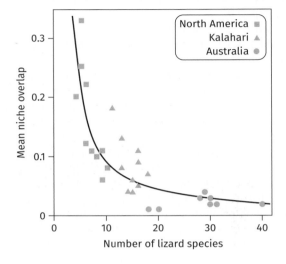

Reproduced with permission from Rappoldt, C. and Hogeweg, P. Niche Packing and Number of Species by C. Rappoldt and P. Hogeweg. *The American Naturalist* Vol. 116, No. 4 (Oct., 1980), pp. 480-492. Copyright © 1867, CCC Republication. www.jstor.org/stable/2460441

© Anthony Short

Another possible issue is the complex competitive interactions that can occur in multi-species communities. Competition could occur in intransitive chains where A outcompetes B and B outcompetes C but C outcompetes A: they are the ecological analogue of the rock-paper-scissors game. Theoretically this could promote biodiversity, but no-one has any idea whether it actually does.

Predators and prey

The other major interaction is predation: does this limit whether a species is regularly present in a habitat? Yes, it does. The very first experimental study that manipulated the presence of a predator showed this dramatically (see Figure 5.9). This was a study of an inter-tidal community on the rocky shores of the Pacific NorthWest of the USA, where the top predator was a starfish *Pisaster*. Its favourite food was the California mussel, *Mytilus*, the dominant competitor for space on the rocks. There was another predator, the dogwhelk *Thais*, but its rather more restricted diet consisted mostly of one barnacle species, with some mussels as well. When Robert Paine constructed cages and removed all the starfish from a particular area, there was a dramatic result. With the starfish gone, the mussels crowded out nearly all the other species in the system—the chitons, limpets, and almost all the barnacles. What was left was a species-poor community compared with what was there before.

The interpretation of this experiment is very important. Some, but not all, predators prevent competition from acting strongly enough to exclude some species. Such predators are called 'keystone' predators—we have already met some (e.g. the wolf) in the concept of the trophic cascade. Keystone predators maintain the species richness of the community by preventing the competitive dominants from excluding the competitively inferior species. They produce what is called top-down control, where species higher up in trophic level control those lower down. The opposite to this is bottom-up control, where the availability of resources (and hence competition) controls populations.

Figure 5.9 The impact of a keystone predator in maintaining the diversity of an inter-tidal food web. The numbers refer to the numbers of species of each taxon. The arrows depict feeding relationships.

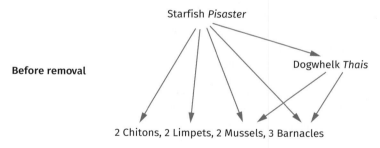

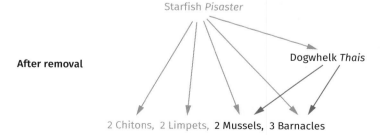

Now we have considered the main issues that influence the composition of local communities. In the assembly of such a community, there has to be a regional pool of species from which to draw colonists for a local habitat, and that habitat has to be accessible and reachable by a colonist. There has to be 'space' in the community to accommodate a new species, either because the community is not saturated, or because space is created by a keystone predator.

Regional processes

The dominant processes influencing species richness at a regional scale and over long time scales are different from those influencing populations at a local scale. The key processes at a regional scale are speciation, extinction, and colonization. A 'region' can be anything from an area encompassing several very different habitats (hundreds or thousands of hectares) up to entire countries and even continents. Six sets of factors have been studied for their influence on the species richness of these kinds of areas:

- climate and plant productivity (net primary productivity);
- the heterogeneity of the landscape;
- soil conditions and nutrients;
- area;
- biotic interactions;
- dispersal and evolutionary history.

There are definitely strong correlations between climate and productivity, and between productivity and species richness. The number of tree species per 2.5-degree square in North America is strongly related to the net primary productivity of the square, and the same relationship occurs in Europe too. Productivity is consistently related to the availability of energy including light, temperature, potential evapotranspiration, and water availability. As a general rule, species richness increases with heat but only when there is enough water. After a certain point, there is enough energy but water limitation takes over.

In different regions the balance differs. Thus in the tropics and other hot places, water limitation dominates, whilst in higher latitudes the temperature is equally or more important than water availability. And as one might expect, the species richness of ectothermic animals is most closely linked to temperature, whilst that of endotherms is much more associated with net primary productivity.

Speciation

Why is there a relationship between the number of species and energy? Since plants use the Sun's energy to produce biomass, and all animals ultimately depend on plant biomass, you might think it obvious that simple biomass and energy input should be related, but why should we have more species as well as biomass? Many researchers have tackled this question, but there never seems to be a clear answer. One of the most popular ideas is that the rate of speciation is related to the generation time of organisms—thus slowly growing species only rarely produce new species, whilst those with rapid population increases produce new species much more quickly. If true, then the rate of speciation would then be related to temperature, and this would generate the observed patterns.

Overall, energy input and water limitation, plus the variability of the topography, accounts for about 70% of the patterns of global plant species richness. This points to a very successful prediction, leaving relatively little for other factors to explain. The topography effect is the heterogeneity of the landscape, and probably indicates increased variety of habitats and resources, the existence of refuges from harsh environmental conditions, and a diverse and patchy set of habitats to encourage speciation.

Isolation is an additional factor—dispersing to an isolated island is often the spur to adaptive radiation, i.e. speciation to fill all the possible niches that are currently empty. There are many spectacular examples of this process—look up the honeycreepers (birds) or silverswords (plants) of the Hawaiian Islands, Darwin's finches of the Galapagos, or the vanga-shrikes (birds) or chameleons of Madagascar, to name just a few.

Extinction

We now turn to study the process of extinction, this time not of a population from an individual patch, but of a species from the Earth. Most species are restricted to particular regions, so we are discussing the regional process of extinction. What kinds of species die out, and at what rate?

We know quite a lot about the kinds of species that become rare and then extinct. There are in fact seven different types of rarity, defined a long time ago by Deborah Rabinowitz. She classified organisms along three axes, each with two categories: there was *distribution* (restricted or widespread), *habitat* (specialized or generalized) and *local population size* (either everywhere rare, or somewhere common). Her point was that there were several ways to be rare, but all of them entail a heightened risk of extinction. Thus a species in category 7 of Table 5.1, is widespread in distribution, generalized in its habitat requirements, yet everywhere where it occurs it is in low numbers. Most of the world's large carnivores are like this, and they are disproportionately endangered.

In fact, there is a syndrome of traits possessed by rare organisms, listed in Table 5.2. They tend to be large predators that have low reproductive effort and poor dispersal, and require particular habitats and food. Often they lack a lot of genetic variation, which makes them more susceptible to disease and less capable of adjusting to changing conditions.

Of course, we all die in the end, and the same is true of species. In the long run, they are all doomed to extinction. Ecologists see a cycle in the history of an individual species, called the taxon cycle. A species first

Table 5.1 Rabinowitz's seven forms of rarity

Rarity	Distribution	Habitat	Local N
1	restricted	specialized	everywhere rare
2	restricted	specialized	somewhere common
3	restricted	generalized	everywhere rare
4	restricted	generalized	somewhere common
5	widespread	specialized	everywhere rare
6	widespread	specialized	somewhere common
7	widespread	generalized	everywhere rare
not rare	widespread	generalized	everywhere common

evolves because of a combination of chance (isolation from the parent species) and opportunity—there is a gap in the 'market', niche space, that can be filled. Once evolved, there often follows an expansion phase where it fills all the geographical space that it can via colonization, and then often a more sedentary phase when movement and hence colonization become less advantageous. Local populations begin to diverge genetically, and some die out, leaving a patchy distribution of endemic subspecies. The end point of this process of differentiation and local extinction is the occurrence of endemic species with very small distributions, coupled with a lot of patches that lack populations—creating more opportunities for other colonizing species.

Thus many of the rare species on the Earth today must be regarded as species at the end of this cycle, basically hanging around until chance creates the conditions that drive the last population extinct. A typical example is the Blue Antelope *Hippotragus leucophaeus* of southern Africa. It

Table 5.2 Traits associated with rarity

Trait	Rare	Common
size	large	small
trophic level	high	low
reproductive effort	low	high
dispersal ability	poor	good
genetic variability	uniform	diverse
competitive ability	poor	strong
resource use	specialist	generalist
habitat use	specialist	generalist
breeding system	asexual, selfing	sexual
sexual selection	strong	weak

became extinct in about 1800, and is probably the only large mammal to die out in southern Africa in the last 350 years (assuming that the extinct Quagga was a subspecies of the Plains Zebra *Equus quagga*, rather than a true species), and the first African large-mammal extinction in historical times. Scientific data were very few before 1800, so this species died out before it could be studied alive, hence the causes of its disappearance are not well understood. It was first recorded in 1719, first described in 1766 by Peter Simon Pallas, and was gone only thirty-odd years later.

A specialist grazer, the Blue Antelope was distributed across the modern state of South Africa in the Pleistocene up to 10 000 years ago, but as the Cape grasslands contracted in area, so did its distribution. About 1600 years ago the first domestic sheep and then cattle arrived, which presumably competed with the Blue Antelope for grass. It ended up with a single population estimated at only 370 animals in a very small region of about 40×100 km in the Western Cape of South Africa. Although finished off by the arrival of Europeans with guns, the population probably only had about 100 breeding individuals—far too low to maintain genetic variation and hence adaptability, and also too low to resist the vagaries of random year-to-year variation in environmental conditions. Such small populations are vulnerable to random extinction, so this species was probably well on its way out.

The giant Steller's Sea Cow (*Hydrodamalis gigas*) is another example. Up to nine metres long and ten tonnes in weight, it was again finished off by Europeans with guns less than 30 years after its discovery. Like the Blue Antelope, it had greatly contracted in distribution over the previous few thousand years, and was probably on the verge of extinction anyway.

We can do a kind of back-of-envelope calculation to see what the rate of this kind of extinction should be. The best estimate of the number of species on Earth is that there are somewhere between 5 and 10 million species. Since we have names for only 1.5 million of these, there is clearly quite a way to go in discovering the rest of them!

If we collate all the data from the fossil record, we can estimate the typical length of time that a species lasts before disappearing for good. Conveniently, this turns out to be between 5 and 10 million years. Thus if this is an accurate way of calculating the 'normal' rate of extinction, then about one species dies out somewhere on Earth every year. This calculation was first done in 1995, and since then we have refined the ways of checking whether it is correct—various lines of evidence suggest it is about ten times too high. Thus we now think that natural extinction rates must be of the order of one species dying out every ten years.

Superimposed on this average 'background' rate of extinction are at least five mass extinction events in the Earth's history. The mass dying-off of entire faunas gave rise to the separation of the Earth's history into the geological periods such as the Triassic and the Permian. The biggest of these mass extinctions—the end of the Permian—was a truly spectacular event, involving the disappearance of 95% of all terrestrial species on the planet. To achieve this, probably 99% of all individuals died.

The evolution of modern humans entailed a dramatic increase in extinction rates, as we shall see. The rates predicted in the future are much higher still, which is why so many people are so worried.

Dispersal, filtering, and mass effects

The link between regional and local species richnesses is all about the processes of colonization and invasibility.

- Is a species able to disperse to habitat X? (= dispersal limitation)
- Is habitat X suitable for the species? (= environmental filtering)
- Does the community already there allow the invasion to take place? (= biotic filtering)
- Are populations that would die out left to themselves, kept going in habitat X purely by immigration of new individuals? (= mass effects, the diffusion of individuals)

All species are limited by dispersal to some extent—no species can get everywhere. There are extremely few species that occur on every continent, for example. Interestingly, the bracken fern *Pteridium aquilinum* is one of these few: it is native to every continent and many islands too, including Hawaii—the most isolated group of islands in the world. Probably this is partly because of its great age as a species—fossil bracken ferns 55 million years old have been found that seem more or less identical to the plants of today.

The inability of species to reach habitats provides lots of opportunities for other species to avoid competition, and hence coexist regionally. This is easy to show using bacteria—three strains can coexist easily on undisturbed agar plates, but two of them are driven to extinction if you deliberately mix the agar.

An example from bumblebee communities

The way the environment and community filter the available species is best shown by an example. We have chosen one from flower-visiting communities of bumblebees along an elevation gradient in the Alps. Altitude differences create very strong gradients in environmental factors, principally temperature and the length of the growing season (i.e. net primary productivity of the plants), so the influence of the environment on communities can be seen more easily. As you might expect, there are more species visiting plants at lower altitudes, so the abiotic and biotic gradients are aligned with one another.

In bumblebees the species trait that is most important in influencing the diet is the length of the mouthparts, because this dictates whether they can reach the nectar in flowers. There are two major related groups (clades) of bumblebees, one short-tongued and one long-tongued (see Figure 5.10). Short-tongued bumblebees visit many more different flower types than long-tongued species, and are therefore more generalized in their food niches. In colder areas there are fewer plants altogether, and year-to-year variation in flowering is much greater: in such circumstances, generalists that are more able to switch among flower types are at an advantage. Thus there is an altitude gradient in the severity of both abiotic and biotic conditions, which should filter species able to tolerate conditions in any one place. In warmer lowland conditions there are more types of flower and the inter-annual variation is much less, allowing specialists to survive there. The relative stability of lowland conditions should promote stronger competition among bumblebee species by limiting their ecological similarity through tongue length.

Figure 5.10 Bumblebee communities along an altitude gradient. The y-axis is the length of the growing season, hence lowland sites are at the top with the longest growing seasons; the dashed line describes how species richness is on average related to the length of the growing season. The left-hand part is the phylogeny of the relevant species of bumblebee making up the communities: they belong to two tongue-length clades; the 'bar-code' in the middle and the lines from the phylogeny indicate the makeup of each individual bee community–the width of each strip is proportional to the number of species.

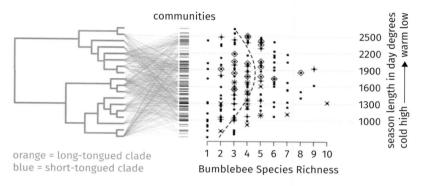

orange = long-tongued clade
blue = short-tongued clade

Reproduced with permission from Global Ecology and Biogeography. Pellissier, L. et al. Phylogenetic relatedness and proboscis length contribute to structuring bumblebee communities in the extremes of abiotic and biotic gradients. A Journal of Macroecology, Volume 22, 577–585. Copyright © 2012, John Wiley and Sons.

The researchers therefore surveyed bumblebee communities along the altitude gradient during the flowering season for two entire years, recording which bumblebee species occurred in each of 149 communities. Figure 5.10 shows the results.

The number of species in the community peaks at intermediate season lengths, declining above and below. Communities from longer growing seasons (lower altitudes) contain bees with more diverse tongue lengths than those from shorter growing seasons (higher altitudes). Thus the severe environments of high altitudes filter out many bumblebee species, and the long-tongued ones in particular, because of the lower availability and predictability of the resources on which they depend. In the more diverse and predictable lower elevations, the number of species that coexist declines because of the more intense competition there (i.e. biotic filtering).

Ecologists think that environmental filtering and dispersal limitation are about as important as each other in influencing the species richness of different communities, and that together they account for as much as half of the spatial variation. This implies that biotic filtering via local interactions such as competition can cause no more than about half of the remaining variation, given that random chance, mass effects, and sampling error also contribute to the data we have. Thus you can see that the regional level is vital to a proper understanding of the composition of local communities.

Implications: the balance of nature

There are some pretty important implications of the issues we have been discussing. One is the very entrenched idea of the 'balance of Nature'. The basic concept here is that, left to her own devices (i.e. without 'wicked unnatural human beings'), Nature (almost always thought of as female) will be in equilibrium, a balance whereby all organisms coexist and maintain themselves. Whole books have been written about this concept, and it pops up everywhere in journalism and popular science, as well as the writings of popular conservation organizations such as Greenpeace.

But Nature is never at any sort of equilibrium. Populations fluctuate all the time, and indeed we know that even very simple models of population dynamics can produce apparently chaotic fluctuations in numbers. Thus even in theory there is not much evidence for any sort of equilibrium or balance. Everything is changing at all time scales, from the fluctuations and cycles of individual populations to the reorganization of entire ecosystems under climate change, natural or human-induced. Populations on patches die out and recolonize, and regionally species become extinct, or colonize new communities, or new ones arise via speciation: all these demonstrate this non-equilibrial view of Nature. Redefining what we mean by 'balance' to describe just species richness rather than abundances does not help. There is no such thing as the balance of Nature: it is a complete myth. And human beings are as much part of Nature as any other species on Earth!

Case study 5.1
Coral reefs

Coral reefs are among the most productive and diverse ecosystems on Earth. Their fish, corals, and other invertebrates come in a dazzling array of colours, shapes, and sizes (Figure A). Corals are invertebrates living in compact colonies of many individual units (**polyps**) that catch plankton and small animals with their tentacles, but they obtain most of their energy and nutrients from single-celled **symbiotic** photosynthetic **zooxanthellae** living in their tissues.

No marine ecosystem has been studied more: over the second half of the twentieth century, gradually a picture built up of a stable system with strong structure and function. Coral mortality was thought to be incredibly rare once the coral had reached a certain size—restricted to infrequent intense local storms. The ecological theory of the time was focused on helping to explain why highly diverse coral-dominated systems were so stable.

Then suddenly in the 1980s, it became obvious that coral reefs were not stable at all. The dominant herbivore of Caribbean reefs, a sea urchin, suffered a catastrophic 90% decline caused by a pathogen, and algal biomass increased rapidly. In fact, there was an entire 'phase shift' from coral to algal dominance, and the fact that such phase shifts were even possible came as a complete surprise. No-one had imagined that such a radical widespread change could be the result of the loss of a single species.

In Australia there was a massive outbreak of a coral predator, the Crown-of-thorns Starfish *Acanthaster planci*, one of the largest of all starfish. Each starfish can eat up to 6 m^2 of living coral per year, so a

CS5.1 Figure A A coral reef—one of the richest ecosystems in the world—but not a stable one!

Rich Carey/Shutterstock.com

population outbreak is a serious issue for a coral reef. There was a worry that the Crown-of-thorns might destroy the Great Barrier Reef entirely. Finally, if those problems were not enough, there was widespread **coral bleaching**, where the coral polyps eject their symbiotic zooxanthellae then usually die: 10% of the world's coral reefs are now dead.

Why was coral not stable? Why were these things happening? Two decades of research has shown that, once again we were looking at the wrong scale—again too small a scale. The drivers of these events were often not within the reef itself, but were caused by its connections with the greater world outside. Global warming is causing the temperature of the sea to rise, and we know now that when corals are too hot they expel the symbiotic zooxanthellae that supply them with most of their food. Chemicals in agricultural runoff are probably responsible for stimulating algal blooms which result in better survival for the larvae of the Crown-of-thorns Starfish. The same factor coupled with the loss of herbivorous fish caused by overfishing also causes algal blooms, switching reefs from a coral-dominated to an alga-dominated state. Almost certainly the dominant driver here is the level of herbivory controlling which state the reef adopts. When corals die from bleaching, the abundance and species richness of herbivorous fish declines, so these factors are all linked.

How will the coral recover? Here again, scale is an issue because fish recruitment turns out to depend not on the coral but on areas of mangrove and seagrass beds that are sometimes a very long way away from the reef. These are the nursery habitats for a wide range of species including many reef fish, and thus the connectivity between reefs and nursery habitats is very important in the recovery process. However, mangroves are currently being deforested at a rate faster than rainforests. Networks of mangrove reserves that are only tens of kilometres apart would be the best solution, but of course this would entail a major upscaling of the current efforts in marine conservation.

 **Pause for thought**

Given what you know about corals, what can the manager of a coral reef reserve do to improve the status of the coral in the reserve?

Chapter summary

- There are four main processes at work in ecological communities:
  - *selective* forces (competition, predation, coevolution) that modify the characteristics (the traits) of populations and species, and/or determine their presence in a community;
  - *ecological drift*, i.e. random chance that affects small populations in particular, that might lead to extinction or immigration events;
  - *speciation*;
  - *dispersal*.
- At large scales the relative importance of these processes is not the same as at small scales.
- The species richness of local communities depends on large-scale species richness via the processes of dispersal, and abiotic and biotic filtering.
- The 'balance of Nature' is a myth—there is no such thing.

Further reading

Wilson EO (1992) *The diversity of life*. Penguin.

Like all EO Wilson's books, this is wonderfully written by an expert in the field. You might also enjoy his 2013 book *Letters to a young scientist*.

Vellend M (2010) Conceptual synthesis in community ecology. *Quarterly Review of Biology* 85(2): 183–206.

This is for conservation biologists, but gives a marvellous overview of what matters in communities.

Levin SA (ed) (2009) *Princeton guide to ecology*. Princeton UP. pp. 186–195, 196–201, 264–273, 274–281.

The Princeton guide consists of short (4–8 pages) chapters written by different authors outlining concepts and ideas in ecology and conservation. Some are rather detailed, but they expand and consolidate the ideas presented here.

Discussion questions

5.1   One of the short-term impacts of habitat fragmentation is to make each habitat too small for top predators. What would this loss do to the food webs?

5.2   Do you think humans are natural or unnatural parts of the Earth's ecological system?

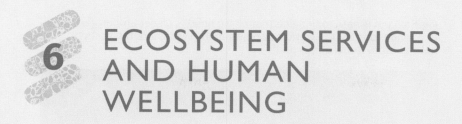

6 ECOSYSTEM SERVICES AND HUMAN WELLBEING

In this chapter we explore the reasons why conservation is so important—and why we cannot afford to ignore the need to do this successfully. The two biggest problems are (a) the sheer numbers of people now living on our planet; and (b) the huge disparities in wealth and lifestyle among and within countries. Those of us living comfortably with no real lack of anything in the more economically developed world create enormous demands on the capacity of the planet to support our lifestyles. We know from simple calculations that it is just not possible for everyone on Earth to have rich first-world lifestyles in a sustainable way—yet it is completely understandable that people all over the world aspire to having plenty of food, education, healthcare, and transport (Figure 6.1). Something has to give.

At the turn of the millennium in the year 2000, the United Nations Secretary at the time, Kofi Annan, called for a stock-taking of the Earth's resources. The result of a huge effort by many scientists all over the world was published in 2005 as the *Millennium Ecosystem Assessment*. The summary volume of 28 pages was entitled *Living beyond our means,* which says it all! The assessment was intended to guide governments in their responsibilities to the environment, outlining the scientific basis for enhancing the conservation and sustainable use of the world's resources and their contribution to human wellbeing. It is this last point that we want to describe here. Just what is it that we all depend upon?

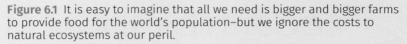

Figure 6.1 It is easy to imagine that all we need is bigger and bigger farms to provide food for the world's population—but we ignore the costs to natural ecosystems at our peril.

Tyler Olson/Shutterstock.com

The idea of ecosystem services

To a modern city dweller living a wealthy lifestyle full of activities and technological advances, it can be easy to gain the impression that we no longer depend on natural systems. We might go for a walk in the countryside to enjoy nature on the weekend, but most of the time we probably operate in the hurly burly of traffic, buildings, illuminated night-life, and all the types of artificiality that we have become used to over the last fifty-odd years. Even rural people often think of natural spaces as unproductive wasteland without value except for its crop-growing or building potential. For the sake of the future of the planet, it is time for everyone to stop thinking like this, and wake up to the reality all around us. These are dangerous illusions—nature is not an optional add-on to our lives. We all rely utterly on the services that nature delivers to us.

Most obviously, we rely on the food that nature provides—an ecosystem service that cannot be taken for granted. This is very obvious when food is harvested directly from wild animals, such as ocean fish. Almost all fisheries are declining in productivity and have been dramatically changed by the process of industrial-scale fishing. For example, one of the most productive fisheries of all time was the Grand Banks area of Newfoundland off the coast of eastern Canada. Huge catches of cod were made and sold around the world. The fishers used to boast that the fish were so abundant fishermen could walk across the sea on their backs. In the seventeenth

Figure 6.2 Atlantic cod had been fished almost to extinction, but bans then strict controls on catches have led to numbers slowly recovering.

(i) Vladimir Wrangel/Shutterstock.com, (ii) Sergio Azenha/Alamy Stock Photo

century, three people in a boat with helpers onshore to dress the catch could collect 30 000 fish in a month. But fishing fleets became ever bigger and more efficient, taking more and more fish all the time (Figure 6.2). The average size of the cod declined from 1.2 m in length in the 1920s to 0.65 m in 2004. The Grand Banks fishery was closed in 1992 because there were virtually no fish left.

The cod have not returned in their original numbers (yet), but their populations are increasing every year and there are hopes that eventually the area may be fished again—sustainably this time. Similarly, the cod fisheries of the North Sea were exhausted in the early years of

Figure 6.3 Without pollinators such as this bee, many of the fruits people enjoy, from the everyday apple to seasonal raspberries, would disappear from our shelves.

Gabriele Maltinti/Shutterstock.com

the twenty-first century. Fishing quotas were dramatically decreased and the number of fishing boats reduced by a half. By 2018, the cod population had grown strongly, and it is now sustainable to take more fish again—but in a very controlled way.

It isn't only wild food like fish that depends on natural systems. Even food grown in what may appear to be the very unnatural conditions of agricultural fields is still the product of the biological processes of nature. Photosynthesis, the replenishment of minerals such as nitrates through natural cycles, the importance of the symbiotic relationships between soil fungi and plant roots, and the role of pollinators such as bees and hoverflies in the production of many fruit crops (Figure 6.3)—these processes and more are part of the ecosystem services on which we all depend.

The impact of the *Millennium Ecosystem Assessment*

A quiet revolution has occurred since the 2005 UN landmark publication, because the report emphasized that the true worth of the services that nature provides are often only fully appreciated when they are lost. The *Assessment* taught us the importance of carrying out a full accounting of the value of an ecosystem service so that it is understood before it is lost. A forest provides timber and fuel, but in a full accounting this is only one third of its true value. We must also include its contribution to air quality by absorbing pollutants, to climate control by absorbing carbon dioxide from the atmosphere into long-lasting biomass, and to flood control by protecting water sources and absorbing rainfall—to say nothing of its recreation value and the benefits to human health and wellbeing of green spaces. Although many of these services are not bought or sold in markets, and hence are easily lost or degraded, nonetheless their value to human societies is high.

Another key point made by the *Assessment* is that the net benefits of retaining a sustainably managed natural habitat are often far greater than its value when converted into an apparently 'productive' landscape. When a full accounting is made, the value of an intact wetland, for example, can easily be almost three times greater than the products of intensive farmland obtained from drained wetland.

The *Assessment* was the first attempt to evaluate the full range of services that people derive from nature on a global scale. It identified 24 measurable services (Table 6.1) and ran a health check on each one. These ecosystem services include things you might assume are obvious—food, fibres for clothing, building materials, and fresh water. They also include some more subtle services—genetic resources for improving our crops and developing medicines, disease regulation, recreation, and ecotourism. Only four of the services were increasing their benefit to humans, whilst 15 were declining and five were stable in most places but declining in some parts of the world.

Table 6.1 The ecosystem services identified by the Millennium Ecosystem Assessment, and their status.

Service	Sub-category	Status	Notes
Provisioning Services			
Food	crops	▲	substantial production increase
	livestock	▲	substantial production increase
	capture fisheries	▼	declining production due to overharvest
	aquaculture	▲	substantial production increase
	wild foods	▼	declining production
Fibre	timber	+/–	forest loss in some regions, growth in others
	cotton, hemp, silk	+/–	declining production of some fibres, growth in others
	wood fuel	▼	declining production
Genetic resources		▼	lost through extinction and crop genetic resource loss
Biochemicals, natural medicines, pharmaceuticals		▼	lost through extinction, overharvest

Fresh water		▼	unsustainable use for drinking, industry, and irrigation; amount of hydro energy unchanged, but dams increase ability to use that energy
Regulating Services			
Air quality regulation		▼	decline in ability of atmosphere to cleanse itself
Climate regulation	global, regional, and local	▲ ▼	net source of carbon sequestration since mid-century preponderance of negative impacts
Water regulation		+/–	varies depending on ecosystem change and location
Erosion regulation		▼	increased soil degradation
Water purification and waste treatment		▼	declining water quality
Disease regulation		+/–	varies depending on ecosystem change
Pest regulation		▼	natural control degraded through pesticide use
Pollination		▼	apparent global decline in abundance of pollinators
Natural hazard regulation		▼	loss of natural buffers (wetlands, mangroves)
Cultural Services			
Spiritual and religious values		▼	rapid decline in sacred groves and species
Aesthetic values		▼	decline in quantity and quality of natural lands
Recreation and eco-tourism		+/–	more areas accessible but many degraded

From the Millennium Ecosystem Assessment: Living beyond our means (2005), p.17

The most obvious services in trouble are the availability of wild fish and fresh water, with strong indications that we have gone past the point where nature can renew their supply sustainably. We are now eating into the capital of the Earth—selling the family silver to fund an unsustainable lifestyle.

The situation is not yet irretrievable. Take fresh water: we only use 10% of the quantity of fresh water that runs into the sea, so you might imagine that there is no problem at all. However, the supply of fresh water is distributed very unevenly across the world, and in many places

does not match demand. For example, Egypt has no rainfall worth speaking of, and no rivers except the Nile, hence all of its water demand must
come either from the flow of the Nile or from ground water sources under
the desert sands. Like other countries in Central Asia and North Africa
with very low rates of water supply, it has a huge 'blue water' demand
(i.e. for surface and ground water). Furthermore, it actually imports vast
quantities of 'virtual water' in the form of food grown with water from
other countries: North African and Middle Eastern countries are some of
the highest net importers of such virtual water. Thus even though overall
the world has enough fresh water, some countries are in serious trouble.

If natural systems were well understood and behaved predictably,
we could calculate exactly how much pressure they can withstand and
still supply the services sustainably. Unfortunately such systems have
a tendency to move from gradual to catastrophic change with little
warning, and the complexity of the interactions among the components
of ecosystems, as we have seen, means that such tipping points are hard to
predict. Once such points have been reached and exceeded, it can be hard
to return ecosystems back to their previous state. So, for example, there
was no warning of the collapse of the Grand Banks cod industry described
above. There was a similarly sudden and dramatic collapse in catches of
the Peruvian anchovy in the early 1970s.

The key role of biodiversity in delivering these ecosystem services is
the main reason for worrying about the loss of biodiversity. The diversity
of nature underpins the services we need, hence the relationship between
diversity and service provision is a vital one. A number of experiments
have manipulated the number of species in communities, and suggest that
the provision of an ecosystem service does indeed depend on species richness, but probably as a decelerating curve (Figure 6.4a). It may also be true
that the variability of the service provision gets smaller (i.e. more predictable) with increasing species richness (Figure 6.4b). If we imagine losing

Figure 6.4 The relationship between (a) the delivery and (b) the predictability of an
ecosystem service (decomposition) and biodiversity (here, species richness of microbes).

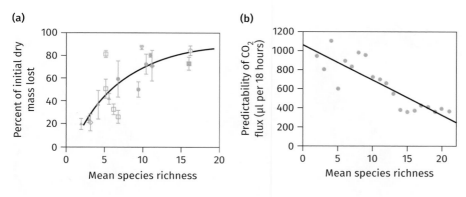

species from the most diverse end of the curves of Figure 6.4, then you can see that initially there seems to be no effect of losing a set of species. After a certain point, however, there are more dramatic and more unpredictable results of losing the same number of species.

The bigger picture 6.1
The story of Costa Rica

Costa Rica is a Central American country which has relatively few resources, yet it has become a beacon of stability in a very unstable area. It has a good healthcare system, levels of education are high, and it is one of the great success stories of conservation on a national scale. Not everything is perfect—it never is—but Costa Rica shows the way in what can be achieved by valuing ecosystem services properly.

In 1900, 89% of Costa Rica was covered in rainforests and cloud forests (tropical forests at altitude usually covered in clouds). By 1989, only 29% of the country was covered in these biodiverse forests. Deforestation was the result of logging for timber, to clear ground to grow cattle to produce cheap beef, and to grow bananas and oil palms.

Then the Costa Rican government made a number of decisions. They did away with their army and used the money to provide free education for all—including

TB 6.1 Figure A The entire population of Costa Rica is educated to value the ecosystem services of their rainforests and their amazing wildlife, such as the green basilisk lizard and the toucan.

© Anthony Short

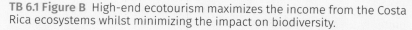

TB 6.1 Figure B High-end ecotourism maximizes the income from the Costa Rica ecosystems whilst minimizing the impact on biodiversity.

© Anthony Short

education about the value of conserving their natural environment (Figure A). They introduced legislation which rewarded land owners for the ecosystem services of their forest and other ecologically valuable land, so the farmers benefited from leaving it undeveloped. They encouraged reforestation and created 20 big National Parks. Devolved government means local people are involved in decisions. Almost 100% of the electricity used in the country is now generated using renewable energy sources, including water and wind.

Over 2 million people visited Costa Rica in 2016. Not only are many jobs linked directly or indirectly to ecotourism, it also generates over 13% of the national income (GDP). But Costa Rica focuses on high-end ecotourism, so the numbers of tourists are limited and the visitors who come to enjoy this magical country cause minimal damage to the ecosystems they come to see (Figure B). Not every country could do this—Costa Rica gains at the moment by being the exception—but it does show what could be achieved.

The government and people of Costa Rica have learned to value the biodiversity of their country. For example, whole villages are now involved in monitoring the beaches where turtles come to nest, saving endangered species rather than collecting the eggs and turtles for food. More than 25% of Costa Rica is now protected, both by National Parks and by private land owners. Costa Rica currently contains around 4% of global biodiversity on only 0.01% of global land mass (Figure C), and has decided that this will continue—an example of a country where ecosystem services are valued at all levels of the community.

TB 6.1 Figure C The small country of Costa Rica has an amazing number of butterfly species (1200), precious biodiversity which must be conserved.

© Anthony Short

❓ Pause for thought

Costa Rica is a Central American country which has avoided many of the major problems experienced by many similar countries. Suggest a list of priorities which other countries might adopt which could potentially both conserve biodiversity, value ecosystem services and increase the standards of living for the whole population.

Valuing ecosystem services

People who make decisions about the environment often have to choose from options that have a wide range of conflicting impacts of a multitude of different kinds. To simplify how to make those decisions, it is a huge advantage if we can convert these impacts into a common standard measurement. This means everyone is comparing like with like, and can understand the basis on which decisions are being made. Converting everything into its value in money is one way of doing this.

In the past, decision-makers have ignored the value of nature, and people arguing for nature conservation have agreed with them because, they claimed, nature was 'priceless'. The trouble with doing this is that by setting the value of nature at zero, we are really saying that nature is worthless rather than priceless. As a consequence, in many decisions that were made, conserving nature was always the least best thing to do because almost anything else that one could do with the land was worth more than zero. Nature does not often have an obvious market price, but 'priceless' is not the same as 'worthless'.

Ecosystem services are not 'free', and the benefits of nature cannot be taken for granted. Thus even though many do not like estimating in monetary terms the value of a beautiful view, for example, it needs to be done in order to set this value against converting natural environments into industry, or building plots, or any other use of the land—and also to calculate the full costs associated with degrading the environment.

Ecosystems can be regarded as environmental assets that, like other capital assets, provide a flow of services over time (Figure 6.5). If the services they provide are consumed in a sustainable way, then the capital remains intact. However, the benefits we currently gain from ecosystems are not being used sustainably, but come at the cost of running down natural capital assets. Some environmental impacts are relatively easy to value. For example, changes in air quality affect (among other things) agricultural production, and changes in production can be valued using the market prices of the day. Changes in health and wellbeing caused by changes in air quality can be costed by increases in visits to doctors and

Figure 6.5 Some of the links between selected ecosystem processes and the goods that people use.

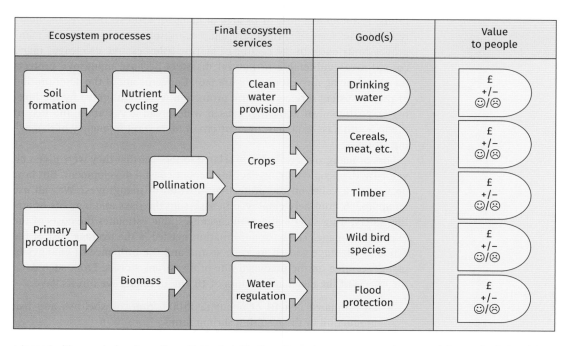

Adapted with permission from Mace, G. M. et al. Biodiversity and ecosystem services: a multilayered relationship. *Trends in Ecology & Evolution*. Volume 27, Issue 1, pp. 19-26, January 01, 2012. Copyright © 2012, Elsevier [https://doi.org/10.1016/j.tree.2011.08.006]

hospital admissions for breathing-related problems. Other impacts are much harder to assess, and a variety of indirect methods is used to try to attach values to them.

How, for example, would you measure the recreational value of a forest to people? Recreational use is one of the most important of the cultural services provided by biodiversity identified by the *Millennium Ecosystem Assessment*. One proposal for measuring its value is called the 'travel cost model' because it measures the travel costs that visitors are prepared to make in order to visit a natural area for recreation. We take these access costs as a proxy for the value of the natural area. Thus, for example, visitors to a forest in Wales were asked where they came from to the forest, and how many times they had visited it in the last 12 months. People living nearby tended to have visited the forest quite a lot, while people from further away had visited only once or twice. Knowing how far each person had come, and by what sort of transport, allowed a calculation of the probable cost of each visit. It turned out that on average the perceived value of each recreational visit was about £12.50, typical of this sort of study. Values from studies elsewhere vary from about 50p up to about £86, but most are less than £5. Multiplying by the average number of visits per year gives an estimate of the annual value of the forest.

Another excellent example concerns estimating the value of the mangrove ecosystem of Thailand. The area of mangroves globally has been declining rapidly, with about 35% of the total area lost between 1980 and 2000. The deforestation was particularly bad in Thailand and other Asian countries as coasts were developed for aquaculture, especially shrimp farms. Thailand lost 56% of its area of mangrove forests in 35 years, nearly all of it because of shrimp farming. Ecologists warned that loss of mangroves contributes to the decline of marine fisheries and also leaves many coastal areas vulnerable to natural disasters because mangrove protects the shoreline against storms (Figure 6.6). Concerns about the deterioration of the ecosystem service of storm protection were highlighted by the Boxing Day tsunami of 2004 that caused widespread devastation and loss of life in South East Asia, part of a pattern of increasing frequency of damaging storms.

Loss of mangroves was clearly related to the fact that they were given no value by the people making decisions about coastal development. But how do we put a value on the protective function of mangroves? We can use an approach, routinely employed in assessing the risks and benefits of airline safety, drug safety, and road casualties, that estimates the costs of the expected damage. The basic information needed is the average cost of the economic damage caused by each storm, and the influence of changes in mangrove area on the predicted rate of disastrous storms. From such data, the effects of the mean annual loss of 18 km^2 of mangrove forests gave:

- an estimated benefit of about £200 000 from the shellfish and fish production of the resulting shrimp farms;
- a much greater estimated loss of about £2.3 million as a result of the increased storm damage caused by the loss of the protective function of mangroves.

Figure 6.6 The economic benefits of shrimp farms seem less impressive once the value of mangroves in preventing storm damage is taken into account.

(i) Alexander Mazurkevich/Shutterstock.com, (ii) xfilephotos/Shutterstock.com

So once we include the value of the ecosystem services of mangroves, is it worth cutting them down and replacing them with shrimp farms? Table 6.2 shows the answer. Shrimp farming produces a lot of toxic waste products that mean that the ponds eventually have to be abandoned and new ones created elsewhere, typically after five years. If we only consider the net income to local coastal communities from the forest products they collect and the fish produced in the mangrove nurseries, then is it only slightly more beneficial to conserve mangroves rather than convert them into shrimp farms. In addition, it is not worth replanting and rehabilitating abandoned shrimp ponds back to mangrove forest because the costs are too high. On the other hand, as soon as the ecosystem service of storm

Table 6.2 The economics of mangrove deforestation.

	Value ha^{-1} ($US)	
	Benefits	**Costs**
Conversion		
shrimp farming for 5 yrs	1078–1220	
mangrove replanting in old shrimp farms		8812–9318
Conservation		
mangrove products	484–584	
nursery linkage to fisheries	708–987	
Total value of products	1192–1571	
storm protection service	8966–10 821	
Total conservation value	10 158–12 392	

Adapted from Barbier EB (2007) Economic Policy Jan 2007: 177–229

protection is included, the total value of the mangrove forest far exceeds its value as a set of shrimp ponds, and it is definitely worth rehabilitating old shrimp ponds.

Thus the value of the ecosystem service is critical to the decision of whether to replant or not. The benefits of storm protection make mangrove rehabilitation an economically feasible option of land use. As we have already seen in Chapter 5, coral reefs and seagrass beds play a role in providing the ecosystem service of fish nurseries, but they also have a storm-protection function. Thus we should really look at the costs and benefits of an integrated mangrove–coral reef–seagrass system in order to understand the full range of consequences of losing them.

Economic arguments are not always helpful to conservation, so we should not rely solely on them. They can encourage overharvesting, particularly if not all the costs and benefits are included in the calculation. Nevertheless, a proper appreciation of the value of the natural world is an important starting point.

The mangrove example demonstrates a further important issue in conservation. The cost of preserving the ecosystem is paid by the local people in the form of giving up the benefits of its alternative use, while the benefits of the ecosystem service are gained by a much wider set of people. This asymmetry in who pays the costs and who benefits represents an important barrier to successful ecosystem sustainability.

The first attempt to put values on global ecosystem services was done in 1997, and the sheer size of the numbers was a shock to many people. It estimated that ecosystems provide at least £33 trillions' worth of services every year, almost twice the global GNP of all the countries of the world. About two-thirds is contributed by marine systems (half of it from coastal ecosystems), as befits its greater surface area, and one third from terrestrial systems, mainly forests (39%) and wetlands (40%). The largest single ecosystem service was considered to be nutrient cycling, at £11.3 trillion.

The numbers bring home the fact that the price we pay for most of our goods and services is too low because it does not include the cost of degrading these ecosystem services. The national accounting systems of the nations of the world are warped—and essentially wrong—because they do not include the depreciation of the natural capital of each country.

The importance of pollinators

We end with a look at one ecosystem service where it is relatively easy to get credible estimates of values: the pollination of crops. Pollination is a key ecosystem service since many flowers require cross-pollination in order to set seeds—and as we know, all life on Earth depends ultimately on plants. 75% of our main food crops, from staples such as potatoes and breadfruit to strawberries and pineapples, are pollinated by insects. We know that the abundance of insects has declined substantially in areas ranging from Europe to China in the late twentieth and early twenty-first centuries. A recent (2017) study of flying insects recorded from nature reserves in Germany found that their numbers had reduced by 75–82% since 1990. This is an extremely serious loss that is already having consequences on the reproduction of rare plants.

Honeybees are one of the best known pollinating insects—and they seem to be under threat the world over. Honeybee colonies are affected by the spread of pests such as mites, by particular pesticides such as neonicotinoids, and by the ageing of beekeepers worldwide (their mean age is more than 60!). But honeybees are not the only pollinators. The more we study crop pollination, the more we realize that wild bees, hoverflies, and other insects do a lot more pollinating than we thought. Thus the maintenance of insect biodiversity and the conservation of pollinators has become an important issue worldwide.

Once again, scale effects are important. Most pollination is done by a few dominant species of insect, but which species are dominant varies across the landscape. Thus a study published in 2018 showed that to achieve 50% pollination of three crops (watermelon, blueberry, and cranberry) in the US in any one farm needs on average 5.5 bee species, while across the entire study area 55 different species are required, because of the differences in the dominant bees of local communities. To achieve 75% pollination, nearly all bee species were needed. Intensive use of the land for agriculture reduces bee diversity and hence crop pollination at both local and landscape scales.

What kinds of values emerge from these studies of pollination services? Of course, it depends on the degree to which a crop depends on insect pollination and the value of the crop. Some of the largest values for pollination services come from the pollination of apple trees in the UK, with values ranging from £9350 to £16 800 per hectare of apple cultivation. The value of strawberry pollination services across the EU is estimated at £10 000 per hectare. For all crops, using a variety of ways of measuring value, recent estimates of the value of pollination services are consistently between £300 and £1300 per hectare. A simple calculation suggests a minimum value of £1.9 billion per year to the UK.

Case study 6.1
The annual value of pollination services to Egypt

Egypt's main arable output covers 70 different plants, including materials used for human clothing (such as cotton), as well as fruits, nuts, vegetables, etc. These plants vary a lot in their dependence on pollinators for successful fruit production and seed set, from full dependence (e.g. watermelons, melons, and custard-apples; Figure A) to total independence (e.g. dates, grapes, maize, and olives).

There are some very useful publications of the Egyptian Ministry of Agriculture that list the market prices of each crop for each year. Thus we can get a good estimate of the value of pollination services using direct market valuation with the Ministry values. While this is straightforward and simple to understand, the true value is only what a buyer will pay, which may well vary across the country, and it also omits other less direct values of pollination services.

The use of each crop varies a lot, and therefore the impact of the loss of pollination services will vary too. Thus some crops are grown for their leaf material, and pollination affects seed production for the next generation of the crop to be grown. For such crops, the impact of pollinators is more long term, but no less serious. For example, a standard fodder crop in Egypt is

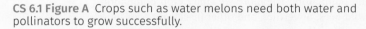

CS 6.1 Figure A Crops such as water melons need both water and pollinators to grow successfully.

matkub2499/Shutterstock.com

barseem (what we call clover, lucerne, or alfalfa), which can be cropped for six years before it needs renewing from seed. However, seed production in barseem is dependent on wild bees because honeybees are especially poor as pollinators for this crop: semi-domesticated solitary bees such as *Megachile rotundata*, on the other hand, do the job very well.

The calculation of the annual value of pollination services to Egypt is dominated, at 46%, by the impact of pollinator loss on barseem, predicted to cause annual losses of about £680 million. There is also a huge economic impact of pollination on melons (£220 million), mangos (£195 million), cotton (£175 million), apples (£135 million), and several other crops. In total the value of pollination services comes out at £1.6 billion per year, representing about 3.3% of Egypt's GDP. This is higher than nearly all the previous estimates from developed European or North American countries. In developing countries like Egypt, pollinator services are almost certainly more significant in that a greater proportion of the human population is maintained by income from agriculture.

The Nile Valley represents an environment with one of the world's longest records of continuous habitat manipulation by humans. Virtually all natural habitats have disappeared, and many insects must have been lost before the advent of modern agriculture, but these losses are largely invisible, as we don't know what was there. Twenty-first-century declines of an already narrowed group of pollinators are likely to have a major impact on the economic and physical health of the population in years to come, unless solutions can be found.

❓ Pause for thought

How do we encourage and conserve pollinating insects? What do you think are the main principles?

⚎ Chapter summary

- Nature provides us with ecosystem services vital to our health and wellbeing. These services have been taken for granted until now.
- There is a relationship between biodiversity and ecosystem service delivery; loss of biodiversity leads to losses in the ecosystem service, both in its level and its reliability.

- Traditionally nature has been valued at zero, hence decision-makers rarely opt for nature conservation rather than using land for other purposes.
- It is vital that we value ecosystem services correctly and fully so that the real cost of their loss is realized.
- Often when the full costs and benefits are assessed, nature conservation turns out to be the most cost-effective and valuable option.

 ## Further reading

Levin SA (ed) (2009) *Princeton guide to ecology*. Princeton UP. pp. 571–583, 584–590, 597–605.

The Princeton guide consists of short (4–8 pages) chapters written by different authors outlining concepts and ideas in ecology and conservation. Some are rather detailed, but they expand and consolidate the ideas presented here.

Millennium Ecosystem Assessment (2005) *Living beyond our means*.

Available from www.millenniumassessment.org/documents/ document.429.aspx.pdf. This very readable executive summary of the assessment is a remarkable overview of the environmental problems that planet Earth faces.

This website gives a useful overview of ecosystem services:

http://uknea.unep-wcmc.org/EcosystemAssessmentConcepts/ EcosystemServices/tabid/103/Default.aspx

 ## Discussion questions

6.1   Some people feel uncomfortable with the whole idea of assessing the value of nature in monetary terms: what do you think?

6.2   What are the different aspects of the full ecosystem service value of a woodland near you?

7 INDIGENOUS PEOPLE AND CONSERVATION

In the final two chapters we turn to the role of human beings in the environments of yesterday and of their role in conservation in the future.

Our early ancestors walked the Earth for thousands of years—you can see some of the evidence in Figure 7.1. It is very easy to blame people for all the environmental problems facing the world. It is also a commonly held idea that early humans, and modern hunter-gatherer tribes, are natural conservationists and work in harmony with their environment. Is this really the case? The advent of modern human beings *Homo sapiens* in Africa 200 000 years ago was clearly important from our point of view, but it was also important for other animals and the environment. What has been the relationship between humans and wildlife over the last 200 000 years? Obviously there has been a major problem during the last 400 years (now called the '**Anthropocene**' to indicate the period heavily influenced by humans), as evidenced by the rapidly increasing extinction rates discussed in Chapter 3, but what about before then? Do humans really differ much from the other animals on the planet, and if so, how do those differences affect the environment for all the species involved? This is what we will be exploring in this chapter.

Figure 7.1 Since the early hominids who left these footprints around 3.5 million years ago, people have left their mark on the Earth. There are just so many of us now!

Images of Africa Photobank/Alamy Stock Photo

Human uniqueness

People have, without doubt, changed the face of the Earth we live on, and had major impacts on the plants and animals with whom we share our planet. Where did people come from, and how have we become such a dominant species so fast? For most of the history of *Homo sapiens*, we have been hunter-gatherers, a society where most or all food is obtained by foraging—collecting plants and animals from the wild. The domestication of animals and plants about 10 000 years ago was a key step in societal development. Societies dominated by domestic animals (pastoralists) or settled agriculture (farmers) have displaced or conquered hunter-gatherers all over the world, leaving hunter-gatherer societies remaining only in marginal habitats such as deserts or tropical forests.

In many parts of the world, people have now changed their environment dramatically, both at a local level with buildings and roads, and globally with our impact on the climate.

We humans like to regard ourselves as unique when compared to the other members of the animal kingdom—but are we really? In what ways do humans differ from their ape ancestors? This can be a controversial question because many religions, non-scientists, and social scientists have views about human origins. Over the last fifty years, however, science has shown that nearly all the ways in which thinkers have imagined humans to be special have turned out not to be true. Those features we thought characterized other animals have been found in humans, and many allegedly human-only characteristics have been found in other, non-human animals. A range of other animal species are perfectly well able to identify individuals of their own species, are self-aware, can deceive one another, show violence towards one another, and can even think of themselves in the position of other individuals. So humans seem to be different in degree rather than in kind—in other words, we have extreme developments of aspects that are present in other animals, rather than being completely different.

So what is left that we can claim is uniquely human? Well, some say not a lot! One view is that most mental features of human beings are a result of having one unique feature—our possession of consciousness. Many developments could be a consequence of this: language, for example, and our ability to see things through someone else's eyes, and our awareness that we will die. Whether animals have consciousness is a key question for which we don't yet have enough evidence, although there is increasing evidence of simple language in some species such as prairie dogs. What 'consciousness' actually is in humans is also open to debate.

What about physical attributes? Among primates we are unique in being mainly furless, and bipedal. There have been some interesting ideas about how this came about. One is based on the fact that nearly all the hairless mammals are aquatic or semi-aquatic, hence it is proposed that a human ancestor was a semi-aquatic diving species around two million years ago. The idea is intriguing, because it links hairlessness with being bipedal, and it accounts for lots of small, otherwise puzzling aspects of modern humans. For example, most aboriginal divers are women, and women have more and differently distributed subcutaneous fat than men. Furthermore, humans who swim a lot in cold water get a peculiar condition called Surfer's Ear, a bone growth within the ear canal that constricts or actually blocks it. The skulls of some fossil humans show this growth, indicating that they probably spent a lot of time diving for food.

An alternative fascinating theory is that we are adapted for endurance running. There is an extraordinary sequence in David Attenborough's TV series *Life of Mammals* from the Kalahari semi-desert, where a San bushman runs down an antelope known as a kudu (see 'Further reading' at the end of the chapter). The kudu is much faster over the short term, but over long distances it overheats and simply runs out of energy and collapses, too exhausted to escape. Over long distances, running on two feet is much more efficient than on four, and overheating is prevented by the lack of hair and the unusually large number of sweat glands in the human skin. Large short-haired mammals do sweat to cool down but they use apocrine glands to do this: humans have lost these almost completely and instead use a completely different set of glands, the eccrine glands. Apocrine glands usually produce odours or milk (in the mammary glands), whereas eccrine glands secrete a clear odourless fluid that is mostly water. The eccrine glands are confined to the pads of the feet in all non-primates, whilst in non-ape primates they are on the hands and feet (and tail if prehensile) and used for improving the grip (via moistening). In the Old World Monkeys the eccrine glands form up to 50% of the skin glands, but uniquely in humans they constitute nearly all of them. This enables humans to minimize overheating, permitting life even in extremely hot conditions.

The Cooking Ape

A strong candidate for a uniquely human trait is the fact that we cook our food. All human cultures cook their food, and the evidence suggests that cooking began at least one million years ago (Figure 7.2). Probably cooking started much earlier, because of the fact that we eat so much meat, yet lack the proper teeth for dealing with it. Meat-eating started at least 2.6 million years ago, about the time of the earliest species of the genus *Homo*, and meat forms 40% or more of the diets of hunter-gatherers today. Yet we lack completely the tearing and cutting teeth of carnivores—instead we have small blunt teeth. Thus now we actually

Figure 7.2 The ability to cook food meant early humans could get a lot more nutritional value from the animals they hunted and the plant food they gathered.

elnavegante/Shutterstock.com

Figure 7.3 The relationship between body size and the proportion of the day spent feeding in primates.

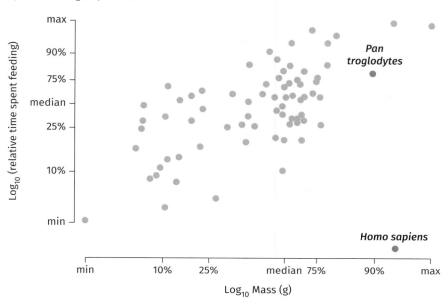

Adapted with permission from Organ, C. et al. Phylogenetic rate shifts in feeding time during the evolution of *Homo*. *PNAS* August 30, 2011 108 (35) 14555-14559. [https://doi.org/10.1073/pnas.1107806108]

require cooked food for survival. We cannot survive as pure carnivores unless our meat contains a very high proportion of fat in the form of rich layers of blubber. Nor can we survive on a raw vegetable diet—people who refuse all cooked food tend to be thin and reproductively impaired even under optimal conditions of eating domesticated fruits and vegetables (with their greatly lowered toxin contents) or lightly processed foods.

Cooking releases nutrients, tenderizes meat, makes starch more digestible, and greatly increases (by at least 50%) the energy available to be absorbed by the gut. Human guts are only two-thirds the length expected for a primate of our size. When we learned how to control fire, and in the process invented cooking, we greatly increased the efficiency of food digestion. This allowed the evolution of a reduced gut and teeth but brought with it a dependence on cooking. The great advantage this gave to humans is shown in Figure 7.3, which plots the proportion of time spent feeding against body size for primates. Large animals tend to be more vegetarian (with some obvious exceptions!) and have to spend disproportionate amounts of time feeding—elephants spend 22 hours per day. Look at the graph for primates. Humans are positioned a staggering distance from the general line relating size with time spent feeding. On average, modern hunter-gatherers spend only 4.7% of their daily activity feeding, whilst a primate our size is predicted to spend 45%. This is a massive advantage, which perhaps generated the time for language and culture to evolve.

Relationships between people and the natural world

In the twenty-first century, we can see the impact of 7.5 billion—and rising—human beings on global ecosystems. Our impact is obvious, from deforestation and global warming (resulting at least in part from burning fossil fuels), to the conservation of threatened habitats and development of seed banks. Our attitudes to the world around us are largely pragmatic, driven by the need to feed huge numbers of people, the demand for minerals and other resources—and the desire to make money. It is easy to think that earlier generations had a different approach—a common idea is that hundreds, thousands, and even millions of years ago, human beings and their ancestors lived in harmony with nature, only taking what they needed and conserving what was around them. Let's explore how realistic that image is, based on the evidence left behind . . .

Hunter-gatherers

Attitudes to the natural world often involve religion. The major world religions, whose antecedents emerged with the development of agriculture and cities, are a far cry from the beliefs of ancient hunter-gatherers. Many modern hunter-gatherer societies have only a very basic concept of 'religion', although a spiritual dimension permeates their normal activities and is continuous with daily life. Simple hunter-gatherer and pastoralist groups are generally much less hierarchical than settled village and city-based societies, and they may hold simpler religious beliefs and participate less in ritual. Women are roughly as powerful and influential as men, even though there is typically a strong division of labour with women doing the gathering and men the hunting. There is usually a strong emphasis on helping kin, and on reciprocity: giving gifts or aid to a person implies the promise of help in return.

All known hunter-gatherer groups have animistic beliefs (see Figure 7.4). This means they believe in the 'animation of all nature', the idea that all natural things have a vital force that can intentionally influence human lives. Other kinds of beliefs are not universal, and analyses suggest that they appeared later in the development of human societies. Animistic beliefs are retained by many societies even today, including nomads and peoples practising shifting agriculture. The 'gods' within the natural world of hunter-gatherers are generally seen as fairly neutral towards humans, but they can be made angry if not appeased. Humans and human lives are regarded as insignificant in the grand scheme.

As an example, we can look at the lives of some of the tribes of New England and eastern Canada in North America before Europeans arrived. Many were originally pure hunter-gatherers, and only later acquired the habit of growing maize, beans, pumpkins, and tobacco. In the summer and autumn they ate wild potato tubers and wild fruits, berries, acorns, and nuts from the trees of the forest. The trees also provided a wealth of materials used in making tools, utensils, and other pieces of equipment. There was usually such a plethora of fish and shellfish that often they did not need to hunt anything. Winter was the most unpredictable season, when people hunted

Figure 7.4 The main features of the belief systems of 33 current hunter-gatherer societies. 'Animism' is defined in the text; 'shamanism' is the belief that certain people can act as intermediaries between humans and the spiritual world; 'high gods' are single all-powerful creator deities; 'active' means acting in human affairs.

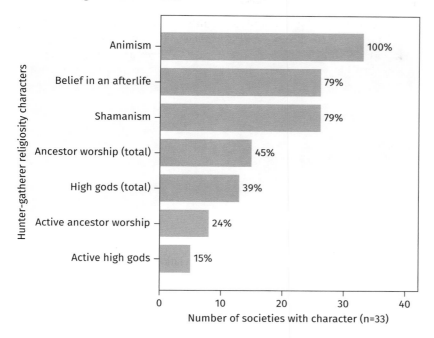

for mammals to supplement their diet—anything from beavers and otters to moose, bears, and caribou. Finally, spring brought a relative abundance of food, from birds and their eggs, to fish and shellfish. Throughout the year, weather conditions made all the difference between famine and plenty.

There was plenty to eat at most times, and these early Americans killed only what they needed, never accumulating stores of anything. Need was the overriding factor. The natural world was the great provider and 'friend'—so long as you were on the right side of it. Hunting was conducted and controlled by spiritual rules determined by the nature of the world around them. Many North American tribes generated rituals and taboos which they applied when animals were killed. For example, in some tribes, when beaver were killed, the bones were never given to the dogs for fear that their sense of smell for beaver would be lost. The bones were not thrown on the fire either, because this would bring down misfortune on the entire tribe; nor were they thrown in the river because the spirit of the bones would inform other beavers, which would then emigrate in order to avoid the same fate. Menstruating women were forbidden to eat beaver lest other beaver decide not to be caught because of this spiritually 'unclean' act. The stated purpose of all these rituals was to propitiate the spiritual keeper of the beaver in order that they continue to supply game to the hunter.

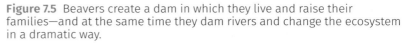

Figure 7.5 Beavers create a dam in which they live and raise their families—and at the same time they dam rivers and change the ecosystem in a dramatic way.

O Brasil que poucos conhecem/Shutterstock.com

Did these practices result in a sustainable relationship between the aboriginal hunter-gatherers of North America and their environment? The waters have been greatly muddied over the decades by the politicization of the relationship between the original inhabitants of North America and their respective governments, and by the undoubted attempts to get rid of the aboriginal tribes, through genocide, displacement, and latterly simply by neglect. There is a certain amount of evidence from early colonists that is relevant.

When colonists from the UK arrived in America, they brought a completely new way of thinking about nature. They regarded everything in the natural world as a commodity—something that could be sold. Perhaps the most famous North American commodity of the time was the beaver pelt. Beaver were incredibly abundant before 1600. They live in family groups in a lodge, which they construct by cutting down trees (Figure 7.5). As a result, these animals alter the surrounding habitat so profoundly that they have been called the archaetypal 'ecosystem engineers'.

Trade in beaver pelts started almost as soon as Europeans came to shore. By the early seventeenth century, tens of thousands were being exported every year. Europeans gave indigenous tribes items that cost almost nothing to manufacture in exchange for beaver pelts, which they could sell at a handsome profit at home in Europe. The First Nations people relished the trade as well, exchanging common pelts worth a trifling amount to them for novel foods and (for them) rare manufactured goods such as copper kettles, metal knives, axes, and cloth. One of them reported to a Jesuit missionary: 'The English have no sense: they give us twenty knives for this one beaver skin'. Complaints of the decimation of beaver numbers soon started to pour in, even as early as the seventeenth century. The material advantages of trading with Europeans thus eclipsed the traditional practices and values animating First Nations hunting.

Originally the indigenous tribes probably regulated their hunting via family 'hunting territories', where a family managed both renewable and non-renewable resources. Many tribes distinguished between mobile animals important for subsistence (such as caribou) and sedentary animals important as commodities for trade with Europeans (such as beaver). The latter were reserved for family use, but anyone could kill the former. But did such territories exist before Europeans arrived? This we do not know. But we do know that afterwards they did not prevent the destruction of many animal populations.

Many people present a vision of humans in the past living 'at one with nature'. However, a very early account from the 1630s does not suggest that the native people in America were conservation-minded. In fact, the opposite is the case: it indicates that when people found a beaver lodge, they 'kill all, great and small, male and female', and that they 'will exterminate the species in this region, as has happened among the Hurons, who have not a single beaver'. There is similar evidence about the overuse of bison by the indigenous North Americans before contact with Europeans. When people find a new way to profit from their environment, they tend to exploit it for maximum benefit, without much thought for future problems.

Figure 7.6 Albrecht Altdorfer's picture of St George and the dragon. Notice the dragon is hardly visible since it merges with the threatening dominance of the great forest.

Interfoto Scans/agefotostock

The Urban Human

Now we are going to consider what happens with the rise of the Urban Human, the agriculturalists whose settled lives led to the creation of permanent villages then cities. Here we see the rise of a different kind of religion, often involving a single all-powerful creator deity actively monitoring what humans do. There is less emphasis on kin-based or reciprocity-based help, and more on helping others at the same level in the hierarchy, or 'group'—religion, club, city, or nation.

In urban traditions, people stopped seeing nature as a benign environment where all organisms can take their rightful place if they abide by certain rules. Where villages grew up in clearings in tracts of encroaching forest, or close to menacing marshes, nature was seen as hostile. European paintings and fairy stories are full of the threats and dangers of nature—Figure 7.6 shows an example of this. Similarly, the Muslim tradition of the Middle East dwells on the beauty of cultivated gardens running with fountains rather than wild nature itself—the harsh and unpredictable desert was seen as a threat to life, and even as the home of Satan. Even today, many mainland Egyptians from the green and fertile Nile Valley fear the surrounding untamed desert.

Cultures that see nature as a frightening enemy demand that she (and nature is nearly always depicted as female) needs to be subdued and exploited. In the traditions of many dominant world faiths, such as the related monotheistic faiths of Judaeism, Christianity, and Islam, the natural world was given by God to humans to serve their needs, and this concept still underlies some attitudes towards conservation today. In contrast to the non-exceptionalist view of early hunter-gatherers (that is, one that sees humans as an insignificant element of the natural world), humans are at the centre of the monotheistic world-view as the special creations of God, entitled to use nature for their own purposes.

The origins of conservation

When people saw nature as infinite and hostile, no-one was interested in looking after it. But at the end of the eighteenth century people came up with a new model of the natural world. Instead of seeing nature as a force to be defeated, Romanticism emphasized an emotional response to the beauty of nature, and the glorification of an unspoilt past. It seems to have arisen as a response to the rationality of the Enlightenment and of science, coupled with the destructive power of the Industrial Revolution.

As well as shaping art and literature, Romanticism had a major impact on ideas about conservation, especially in America. For example, in 1845, 28-year-old Henry David Thoreau—an influential American writer and naturalist—went to live in the woods by Walden Pond for two years. He kept a detailed diary and used it as the basis of a book called *Walden, or Life in the Woods*. This book became the foundation stone of the conservation movement in America. It suggested that humans need nature for mental and spiritual regeneration, a refuge from the corruptions of the city life. A key element was that nature could only act as a spiritual resource if it were 'pristine' wilderness, i.e. *untouched by human beings*.

The eventual outcome was the establishment of the world's first National Park, Yellowstone, in 1872. To provide the 'pristine' wilderness that the whole idea of conservation then embodied, the indigenous First Nations peoples were excluded from the Park, despite the fact that they had made seasonal use of the area for 11 000 years, and that some of the Eastern Shoshone had lived there all year round.

British conservation had a completely different origin. It stemmed from colonial experiences, mostly in Africa, where game hunting was a major sport of the aristocracy. Broadly speaking, colonists cleared humans and animals from 'underused' good land for farming, leaving only the poorer land for wildlife and local people. Sport hunting was regarded as a gentlemanly pastime that could not possibly cause suffering and wildlife loss. It was therefore assumed that any environmental damage must be the result of local subsistence hunting, which distracted rural people from doing 'proper work'. When reserves were established to maintain wildlife stocks for sport hunting, local people were once again removed from their traditional lands, and their hunting practices redefined as 'poaching'.

The images of the Earth from space from Apollo 8 (see Figure 1.1 at the beginning of this book) brought home to many people the limitations of the Earth's resources. They gave birth to the idea of the Earth as a fragile ecological system that could be damaged—as first shockingly reported by Rachel Carson in her book about the impact of pesticides, *Silent Spring*, in 1962. Globalization brought a tidal wave of American ideas to the world, including Thoreau's concept of nature as 'wilderness', which dominated the subsequent growth of the international conservation movement. In its most extreme form, the idea is that humans should 'live in harmony with nature', with the subtext of 'like we used to do thousands of years ago'. As in hunter-gatherer communities, humans are once again seen as having no inherent superiority to other species, despite being powerful enough to damage or destroy nature. In this world view the special role of humans—albeit woefully neglected—is as stewards of nature.

The bigger picture 7.1
The myth of living in harmony with nature

The myth of living in harmony with nature is fundamentally just that—a myth. But it is a myth that contains a paradox within it. The evidence shows clearly that the evolution of modern humans 200 000 years ago entailed a dramatic increase in extinction rates, an indication of the efficiency of intelligent cooperative predation. During this so-called 'Pleistocene overkill', humans obliterated 40–70% of the large mammals in Africa. As humans migrated out of Africa about 60 000 years ago into other continents, there were waves of extinction of large mammals everywhere we went: in Australia, 86% of the species were lost; in North America 73% died out; and in South America, 80% were driven to extinction.

BP 7.1 Figure A The extinction of the Shasta ground sloth *Nothrotheriops* in south western North America. The data show the numbers of Shasta ground sloths recorded in suitable sites, orange for sites where the sloth was present, and white where it was absent. The migration of humans is shown as a blue band: the sloths immediately declined after this date, and were extinct 1000 years later.

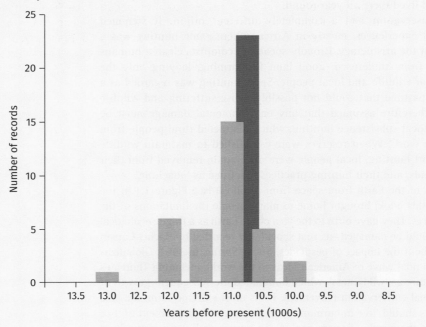

Science History Images/Alamy Stock Photo

A typical example is shown in Figure A—that of the Shasta ground sloth *Nothrotheriops* of the New World. It was at its peak abundance about 11000 years ago, but it declined in frequency immediately humans arrived, and became extinct less than 1000 years later. This pattern was repeated in one large mammal species after another all over the world.

We can see the same effect of the arrival of humans on the Polynesian islands of the Pacific. Almost every island once had at least one endemic species of rail—secretive birds related to coots and moorhens. The arrival of the Polynesians saw about 1000 non-passerine (non-perching) bird species wiped out—that is about 10–15% of the world's total. And all this **before** the Europeans got there and eliminated many more species!

Our best estimates of the background rates of extinction before humans evolved are at least 1000 times lower than after we arrived. The crucial fact is that early *Homo sapiens* did **not** live in harmony with the local wildlife. On the contrary, they exterminated certain types of animal completely. And as soon as they got their hands on more efficient killing technology in the form of guns and metal knives, more species were wiped out—as seen with the beaver in areas of North America.

Researchers have looked at the hunting practices of modern hunter-gatherer peoples to see whether there are any signs of restraint that might indicate conservation thinking. Overwhelmingly the data support the idea that conservation was not on their agenda. Because their social conditions do not permit them to see the wider picture beyond their own environs, they hunt to maximize short-term gain, heedless of its impact on future rewards. The early Jesuits in North America noticed the same thing—food was never stored by tribal people but was eaten immediately, either because they lacked the appropriate technology (pots, etc.), or because everyone was confident that nature would provide more when they needed it. A recent consultation with indigenous peoples reported that even today 'poverty is a foreign concept for peoples who see the Earth as a sacred provider of all the things they need to survive'.

Even if early humans were responsible for widespread overhunting of large mammals, indigenous peoples today, including many hunter-gatherers, cannot feasibly be regarded as currently destructive of their environment because:

- whatever environmental destruction humans did impose happened thousands of years ago, creating modern ecosystems;
- population densities of indigenous peoples in the modern world are simply too low to have much destructive effect—especially in comparison with industrialized environmental exploitation that international regulation has been relatively powerless to stop.

Humans have always affected their environment, even in places we might imagine were 'pristine'. A recent study of the deep tropical forests in the Congo Basin of West Africa and in the Amazon of South America failed to find *any* place where the signs of humans were absent. Every site had been used for shifting agriculture, or had been burned, or otherwise used at some time. The same is true in industrialized countries such as the UK. All of the 'wild' countryside people want to protect is in fact the product of human activities. If we stop practices such as grazing or burning, precious (albeit relatively recently constructed) ecosystems will be lost during successional processes. Human hands have moulded the natural world everywhere.

The myth of the 'Noble Savage', the notion that indigenous peoples live 'in harmony with nature' as they have done for thousands of years, looks untenable in the light of the evidence. Hunter-gatherers were certainly 'destructive', if by that we mean that they changed their abiotic and biotic environment. But all top predators can have that effect, as we have seen in the idea of the trophic cascade. Humans are highly effective as predators because they cooperate together in achieving aims—something few animals do, especially not large predators (lion prides are an obvious exception). Although their effects on large mammals were formerly dramatic, hunter-gatherers now occur at densities too low to change much more of their environment. Equally, there is little evidence that modern urban humans are uniquely destructive of nature. The real cause of the negative environmental impact of modern humans is their numbers—never before has there been a large, highly effective predator at the current population density of human beings. The intensity of environmental manipulation required to support this density, coupled with modern global economic systems that promote environmental exploitation in the service of over-consumption, have resulted in the unprecedented levels of environmental damage we see today.

❓ Pause for thought

When asked, a substantial number of people say that their ideal lifestyle would be living on a smallholding, growing their own food, and keeping a few livestock: discuss how this reflects the myth of the Noble Savage.

Pragmatic conservation

In what circumstances do indigenous people behave like conservationists? The answer is, whenever it is clearly in their interests to act in a sustainable way. Many of the world's poor are faced with fundamental dilemmas: it may be more rational in the short term to exploit a resource, even to extinction, if it secures the immediate survival of themselves and their children. With our full stomachs, cars, medicine, and high standard of living, we cannot expect poor people in developing countries to be conservationists just because we tell them it's a good thing. There have to be incentives and mechanisms that encourage long-term sustainability and support livelihoods in non-damaging ways. What kind of mechanisms might these be? A number of key issues need to be addressed in areas where indigenous peoples still live:

- *Access rights to the resources of the landscape*: If these are open to everyone, then cashing them all in for money might be a rational response. Outsiders need to be controlled or excluded: in West Africa, for example, long-term residents use resources much more sustainably than recently arrived immigrants. An alternative way of putting this is to say that the rights of indigenous people must be acknowledged and respected if they are to be able to live sustainably. If they are not, as in the excluded peoples surrounding the African parks established under colonial rule, then local people become antagonistic to the whole project of conservation and seek to undermine it.

- *Control*: Local people must be able to control the regulation of the resources. If they can, then they have every reason to detect those who try to cheat the system and take more than their fair share. Detecting and penalizing cheats is one essential mechanism of enforcing cooperation in the sustainable use of the environment, and is often built into the customary law that governs tribal and other societies.

- *Long-term use*: There must be a mechanism for passing resources from one generation to another, either within a family or a group of families or a tribe. Whatever the level of inheritance, the long-term benefits of conserving the environment must be gained by the descendants of those who act in a sustainable manner.

Where indigenous people are found to be regulating their resources, it is often an indication that all these preconditions are in place. Then family groups can have territories, and tribes can regulate the use of areas by leaving some untouched for periods of time. More research is needed to understand how such mechanisms work, and what influences whether they are present or not. We also need to understand how to integrate such methods with modern lifestyles.

Case study 7.1
The Jebeliya Bedouin of South Sinai—the relationship between local people and conservation

This case study involves our own decades-long research programme in South Sinai. The material draws heavily on our research findings.

For centuries, South Sinai's indigenous population consisted just of Bedouin, left mostly to their own devices by successive rulers and governments. 'Bedouin' is the anglicized version of 'Bedu'—'people of the desert' in Arabic. The approximately 40 000 Bedouin in South Sinai are divided into eight tribes distributed in the various parts of the region (Figure A).

The modern world arrived with a bang in South Sinai in the form of the Israeli invasion during the Six-Day War (1967) and subsequent occupation.

CS 7.1 Figure A The tribal territories of the Bedouin of South Sinai. 1: Jebeliya; 2: Awlaad Sa'iid; 3: Sawalha; 4: Gararsha; 5: Aleygat; 6: Mzeina; 7: Laheiwaet; 8: Tarabin. The Protectorate boundary is marked in dark blue.

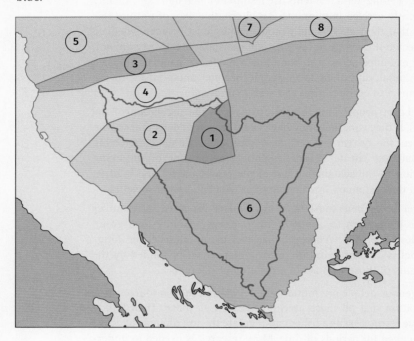

This was followed by the return of the Egyptian administration in 1982 with a new interest in reinforcing its control in Sinai. Both effectively treated Sinai as a colony and the indigenous Bedouin residents as subordinate. Successive Egyptian governments have largely ignored their needs. The jurisdictions have thus produced conflict with the Bedouin, who understandably feel undervalued and neglected. South Sinai Bedouin are very poor: well over half live on less than US$1 per person per day (the UN definition of absolute poverty). In remote regions many still have no schools, healthcare, electricity, or water.

Their nomadic or semi-nomadic lifestyle has gradually declined as the tribes have settled into fixed dwellings, by choice or otherwise, and have become more integrated with settled communities. This loss of traditional identity has been exacerbated by the low esteem in which nomadism as a way of life is held by political decision-makers all across North Africa and the Middle East, where they are universally—but probably unjustifiably—blamed for overgrazing. The older Bedouin generation are the last whose lives were formed within the context of traditional culture.

The Bedouin are traditionally called nomadic pastoralists, but in South Sinai they practise transhumance rather than true nomadism—this means they move seasonally between the same winter and summer areas with their herds of goats and sheep. They also fish, grow dates, create gardens, make charcoal—and sometimes are involved in smuggling. They have always used their unrivalled knowledge of the landscape and environment to guide travellers. Recently they worked as menial labourers under the Israelis, then became involved in tourism, both in the town of St Katherine and also on the coast at Sharm El Sheikh, Dahab, and Nuweiba.

The Bedouin are broadly egalitarian in their outlook—generosity is more important to them than hoarding wealth, so they are suspicious of people who appear wealthier than others. They do have positions of leadership, electing them *(sheikhs* and headmen, or *omdas)* for a fixed term. They are superb poets, using poetry in everyday life to express their feelings about personal or political issues. Bedouin culture developed without being written down, so poems are memorized and recited again and again. They have a well-developed tribal law, which created and enacted a legal system without the need for a government. Virtually every aspect of their culture is an adaptation to the arid environment: their beliefs, social behaviour, law, poetry, diet, and dress have all developed as answers to the challenges that desert life creates.

The Jebeliya tribe (see Figure B) has about 4000 members living mostly in the high mountains around the town of St Katherine, the only non-coastal town of South Sinai. 'Jebeliya' means 'people of the mountains'. In 1996, the entire area of the southern mountains of Sinai, about 4350 km², was declared

CS 7.1 Figure B Two Jebeliya men: on the left, Mohammed Khedr Al-Jebaali; on the right, Hussein Saleh Musa Abu-Meghanim (both shown with their permission; it is not permitted to show photos of women).

© Hilary Gilbert

a Protected Area, completely enclosing the Jebeliya tribal lands within its core. It is Egypt's most important area for biodiversity, representing only 1% of the land area yet containing 25% of Egypt's plants, 67% of the butterflies, and 36% of the mammals. Initially an enlightened approach to conservation meant that, unlike the native Americans of Yellowstone, they were not thrown out of the park. Instead the park's mission was to incorporate the people into its conservation practices.

Both inside and outside the town, the Jebeliya live in small settlements of 4–5 families. There are about 40 such settlements in the St Katherine area, with others scattered across the high mountains. While the men do whatever paid work is available, nowadays the women hardly ever leave their settlements: unmarried or elderly women traditionally took the herds out for grazing, but this has become less common because the arrival of tourism makes it more likely that women might encounter a stranger. Each extended family has a variable number of goats and sheep, occasionally a few hens, and the better-off have 1–2 camels—some also have donkeys to carry water and baggage. Pastoralists everywhere have found that about 40–50 goats and sheep are required in order to provide enough to supply a family's needs, but most Jebeliya families settled in the town have far fewer than this (typically 5–6) because of the risk of localized overgrazing—there simply is not enough grazing within a day's walk of the town centre.

Away from St Katherine, the Jebeliya have a sophisticated way of sustaining the grazing over the long term in a system called *hilf*. This is essentially an agreement backed by tribal law to rotate areas of fallow land, that is to select particular wadis (dried-up river valleys) and deliberately not use them for grazing for a specified period of time, or until the plants have recovered to a specified height. Sometimes only certain animals are allowed to graze. *Hilf* is also applied to the cutting of live trees and the collection of dead wood.

A local person is appointed as a monitor, whose job it is to see that no accidental or deliberate flouting of the agreement occurs. The breaking of *hilf* has serious consequences in tribal law. A person sending animals to graze in a *hilf* area would either be fined an enormous fine, or lose the animals to the person reporting the incident. Islamic principles support *hilf* but prevent its over-zealous application that might deny food or fodder to hungry people. The rules also apply on a large scale among tribes, so (for example) during periods of drought, areas with water are accessible to members of other tribes from drought-stricken lands.

Traditional small-scale orchard gardening (Figure C) is the other typical traditional occupation of the Jebeliya. These orchards are found in walled gardens in the valleys of the high mountains, supplied with water from wells. Some gardens have been cultivated for 1500 years: on average each garden is about 2000 m^2 in area, and contains 50 trees. There are more than 600 of these gardens within the core of the Protectorate. Most importantly for conservation, the gardens act as 'islands' of biodiversity in the sea of relatively species-poor desert habitat that surrounds them. The unpredictable rainfall seeps into the bedrock, to be harvested via wells and dams for use in orchard gardens, allowing the Bedouin to cultivate a wide range of trees and crops throughout the year. Rainwater harvesting techniques such as these are known to improve crop yields and enhance food security in arid regions.

CS 7.1 Figure C A Jebeliya garden with the apricot tree in full bloom in April.

© Hilary Gilbert

In the gardens, in addition to the orchard trees, the Bedouin also grow culturally important but minor crops such as fennel and mint. These elements have a dramatic positive effect upon the structure of plant-pollinator visitation networks, supplementing the resources provided by wild flowers, and maintaining the pollinator community through the hot summers. In early spring the presence of wildflowers has a positive effect upon pollination services to the primary crop (almond), by attracting higher densities of wild pollinators into the gardens and facilitating enhanced fruit set. The higher abundance of resources within the gardens increases the variety and density of birds in the region, and are particularly important for spring and autumn European migrants, providing an important stop-over for numerous small migrant bird species.

Rain-fed irrigated agriculture in arid environments has the potential to increase biodiversity above that found in the unmanaged environment. In St Katherine traditional Bedouin practices enhance wildlife within the Protectorate, and thus initiatives to fund and support gardeners should be encouraged. Rainwater harvesting in arid regions offers a low-cost strategy for increasing agricultural productivity that does not undermine the biodiversity on which it depends.

The value of the gardens to Bedouin wellbeing

Compared to World Health Organization standards, Bedouin children are well below average weight for height, with 13% of them malnourished according to the standard definition, and 6% of them extremely malnourished. Figures for the general Egyptian population are 7% and 1% respectively, confirming the position of Bedouin as among the poorest and most marginalized of all Egyptian citizens. The degree of malnourishment becomes more and more pronounced as the children get older, but improves if the family has livestock, especially if these feed on natural vegetation rather than purchased fodder.

The largest impact on child weight-for-age is of course the amount of household income to buy food—the more the family has, the better their children's health is. A close second in impact is the ability of women to make a cash contribution to the family economy (via paid work, selling produce, or sale of handicrafts to tourists—an option which has now vanished with the recent collapse of tourism in Sinai following terrorist incidents mostly in the North). A third impact is the source of the family's water: the possession of their own well makes a positive impact, whilst being forced to obtain their water by buying it from a water tanker makes a negative impact on weight-for-age (probably because of sickness arising from the poor quality of commercial water). The contribution women can make 'in kind' (i.e. by tending goats for milk, or from garden produce for the family table) also has a clear positive effect, this time on the weight of infants (1–2 years of age).

Gardens are owned and worked by the men, but the produce is often owned by the women. Tribal law has a lot of rules to protect the people who make the huge investment that creating a new garden represents—traditionally the resources needed to survive for an entire year. If the garden owner is not present, then no entry is allowed: entering a garden without permission is regarded as an insult and a violation of the privacy of the family.

If this law is broken, there are extremely serious consequences. If theft is involved, then every individual decision is judged as deliberate and significant—for example, each step from the public trail to the garden wall is counted; mounting the wall, entering the garden, walking to the tree, reaching for the fruit, then leaving the garden are all counted individually and each component costs a camel. The judge would convert this into money—at current rates a fine of six camels would be between 20000 and 60000 Egyptian pounds (around £1000–2500). At the time of writing, most Bedouin live on less than 75p per head per day, so these sums are astronomical. The penalties may seem draconian, but the value of a garden is immense. As is normal in Arab culture, the law allows for exceptions when a traveller is in danger of dying of lack of water or food.

Conclusion

The traditional livelihoods of herding and orchard gardening benefit the growth and development of Bedouin children, and should therefore be encouraged and supported by policymakers concerned with addressing Bedouin wellbeing. The relationship of the Bedouin with their livestock and their gardens, and the natural environment in general, is fundamental to Bedouin culture and identity. Bedouin traditions thus enhance and maintain Bedouin wellbeing, not only culturally in terms of their self-image and feelings of self-worth, but also directly in improving the health and growth of their children.

The extreme conditions under which the Jebeliya live, and the prejudices they suffer, are typical for indigenous people all over the world. They are almost certainly not destructive of their environment, as so often asserted. On the contrary, their traditional law of *hilf* has preserved the vegetation upon which their goats depend. Moreover, studies in South Sinai and similar habitats elsewhere suggest that plants grazed over centuries actually adapt to being grazed and improve when they are. As humans have done everywhere, Bedouin have modified their environment over thousands of years, but its long-term sustainability in its current state is (probably) helped by the rules of tribal law. Trying to imagine what the environment would be like in its 'pristine' state without human beings would be a fruitless task.

 ## Pause for thought

What evidence would you need to be convinced that in some environments indigenous people are good conservationists?

 ## Chapter summary

- Since they first evolved, humans have both modified and been part of the environment. Evidence that hunter-gatherers live 'in harmony with nature' is poor. Evidence that urban humans are uniquely destructive is also poor.
- Attitudes towards nature were universally animistic in hunter-gatherers, but changed in urban humans, who usually thought of nature as a threat and as dangerous. In Western thought the Romantic movement produced the idea of conservation as a way of keeping 'pristine' nature as a spiritual resource for humans.
- Studies of modern indigenous peoples show they often have mechanisms for sustaining their environment that can easily be lost when western-style development arrives.

 ## Further reading

Davis DK (2016) *The arid lands: history, power, knowledge.* **MIT Press.**
An important history of the myths held by Europeans about the environments and peoples of arid ecosystems.

Diamond J (1991) *The third chimpanzee.* Hutchinson.
  A bit out of date, but a very readable paperback on human biology and behaviour as animals and mammals.

Krech S (1999) *The ecological Indian: myth and history.* WW Norton, New York.
  An easy-to-read discussion of the First Nation peoples of North America and their relationship with their environment before and after contact with Europeans.

Ridley M (1996) *The origins of virtue.* Penguin.
  A discussion of what can be learned from viewing human behaviour from an evolutionary standpoint, written by someone who knows what he is talking about.

Robbins P (2012) *Political Ecology.* 2nd edition. Blackwell, Oxford.

Zalat SM & Gilbert F (2018) *Gardens in a sacred landscape.* Amazon.
  An account of the Jebeliya Bedouin gardens and their cultural significance.

Life of Mammals sequence: *www.youtube.com/watch?v=826HMLoiE_o*
  The sequence where a San bushman runs down an eland to exhaustion.

 Discussion questions

7.1   Conservation could be very successful in many areas of the world where the traditional lifestyle of indigenous people plays an important part in maintaining the biodiversity of the ecosystem. But can such people choose their traditional way of life over a more modern lifestyle? Discuss the arguments for and against building conservation programmes based on maintaining both traditional practices and biodiversity, but which may work against people having access to education and modern health care.

7.2   Human beings are inherently destructive. Discuss.

8 CONSERVATION STRATEGIES

This final chapter draws together the threads of what we have been discussing throughout the book. We invite you to think about what side you are on in the arguments that we outline, because there is a great debate taking place in conservation science about the relationship between conservation and indigenous people.

On one side are biologists who think that parks should be people-free and defended against local people who exploit biodiversity. On the other are biologists and social scientists who think local people are the key to conservation success. The latter group thinks that conservation cannot simply be imposed from the outside—it needs the expertise and the good will of local people to succeed.

We emphasize again here the large scale at which conservation needs to happen, which in turn means embedding conservation thinking in the whole economic and political landscape, rather than hiving conservation off to be purely the concern of Protected Areas (Figure 8.1).

Figure 8.1 Conservation and biodiversity are issues in all habitats, from UK bogs and heath lands to deserts as far apart as Africa and the US.

(a) Martin Fowler/Shutterstock.com, (b) Adwo/Shutterstock.com

Finally, we reiterate the fact that the human race has no choice about conservation if it is to survive, because we all depend on the services that nature provides. The big question which must now be asked is how these goals can be delivered.

The great conservation debate

We often take the science we are presented with at school or university, or on the TV or internet, as fact. But behind much of the science we meet are sets of hidden assumptions, great underlying themes that often go unquestioned or unarticulated by the majority of people working in the field. These paradigms underpin much of our activity, and yet until we stop and think, they can go unrecognized for a very long time.

For example, a mechanical approach dominated nineteenth-century physics—the 'Newtonian paradigm': if a physical model of an idea could not be built from metal and string, then it must be wrong. This worked perfectly well until people began to consider the microscale of atoms and electrons. Here, physical models let us down. We needed the idea of random chance and the peculiarities of Einstein's new paradigm of relativity to help us grasp the new 'reality': 'Physics has lost its walls' was one comment at the time. Paradigms are not specific to science—you will find them in every area of life and intellectual thought.

Conservation has its own paradigms. We have already looked at one—Thoreau's Romantic view of pristine people-free nature as a spiritual resource for the renewal of jaded city people. After decolonialization, Westerners were shocked by the attitude of the new leaders of African countries, who stated very clearly in the Arusha declaration of 1961 that wildlife had to pay its way. No airy-fairy spiritual resources for the new regimes—wildlife was an economic resource and they would conserve it only if it were useful. When people are trying to survive, to develop economies, and to educate a population, they do not have the luxury of conservation as the spiritual refreshment of the few.

We see a conflict of paradigms in the Great Debate of conservation (see Table 8.1).

Table 8.1 The Great Conservation Debate.

	Nature protectionists	Social conservationists
Goal of conservation	preservation of biodiversity for its own sake	sustainable human development (hence social justice & poverty alleviation)
Main technique	Protected Areas free of people	people–park partnerships
What works	integrated projects mostly fail; only Protected Areas work	Protected Areas usually fail; success will only happen with the buy-in of local people
Main local threat	poaching by local people	opportunity costs to local people when park created not replaced by alternative ways of benefiting from wildlife
Main local benefit	ecotourism	use of wildlife must be better than all alternatives

Nature protectionists

On one side are the nature protectionists, who think we must conserve nature for aesthetic, moral, or spiritual reasons. Wild pristine nature is beautiful: it is immoral to destroy it, and we need it for spiritual renewal. They are convinced that what we need are more Protected Areas free from the influence of people, where nature can flourish and/or recover. For them, the goal is the protection of biodiversity, and the only tool proven to work is a barricaded reserve, free from people.

The task is described in military terms: a battle to save the planet, a race against time, with Protected Areas as beachheads in the global war against extinction. Indeed, this side of the argument is also called '*fortress conservation*', in which people-free parks and reserves are seen as nature's last stand, her last line of defence against development and eventual destruction. All too often, the local indigenous people are removed from their former lands and deprived of access to their traditional resources. They often live in poverty around the nature reserves and national parks, and are forced to accept the damage caused by large animals coming out of the park (Figure 8.2). Under this paradigm, local people embody the forces of destructiveness threatening the animals and plants of the park by poaching, illegal logging, or grazing. Its promoters tend to be conservation scientists, ecologists, and environmental biologists.

Figure 8.2 For many people in countries such as the UK, elephants are iconic animals to be protected at all costs. If you are a local villager, desperately needing your crops to do well to feed your family, elephants can be the enemy. Nature protectionists would protect the animals regardless of the cost to the people. Social conservationists would look for ways to help the people keep the elephants away or get acceptable compensation for the damage caused.

Signature Message / Shutterstock.com

Social conservationists

The contrary view is made by the social conservationists who advocate various forms of sustainable use of wildlife. Their approach is that social justice and poverty alleviation are priorities that cannot be brushed aside to accommodate wildlife, as they have been so often in the past. This view suggests that conservation should be orientated towards social and economic development. It defines the goal as sustainable development; in other words, improving the quality of human life within the carrying capacity of supporting ecosystems. Its supporters design integrated programmes of conservation and development, using such tools as promoting the sustainable economic use of park resources, and ecotourism, where local people have a substantial input and hence gain, for example by being paid as nature guides. This paradigm suggests that local indigenous people can be allies in conservation if they are appropriately involved in the conservation effort. Its supporters tend to be environmentally orientated social scientists (anthropologists, political ecologists, etc.) and development professionals, but with increasing numbers of conservation biologists as well.

The two sides argue about what actually works (Table 8.1). Social conservationists say that very few Protected Areas are properly funded and staffed, and indeed many are often just parks on paper alone, with inadequate resources devoted to protecting biodiversity. In such cases, sacrificing traditional ways of life in order to protect wildlife cannot be justified. Fortress conservationists point to the general failure of integrated conservation-and-development projects to deliver on either of their goals. It must be said that in general such projects do seem very vulnerable to political interference and corruption. Often they generate only an income supplement for local people, rather than producing lasting new livelihoods that either validate traditional practices or offset the costs of renouncing them in the cause of conservation.

The disagreement about whether parks or integrated development projects 'work' to conserve biodiversity is really a technical issue solvable by data. But the different underlying paradigms are also really important. Just what are the ultimate aims and reasons for conservation? Should the protection of biodiversity be the main goal, or should it be one of the elements of a programme whose main goal is sustainable development?

Fortress conservationists do not deny the need for development and the alleviation of social problems, but they hope that these might be a beneficial by-product of a biodiversity-driven policy of Protected Areas. Social conservationists realize that many indigenous peoples use natural resources both for their subsistence and their cultural survival. They think conservation policy must recognize this and, where traditional practice is held to be damaging, offer acceptable alternative livelihoods. If resource-use restrictions leave people unable to support their families, social conservationists believe that any attempt just to conserve wildlife is doomed to failure.

Learning to live with nature

In fact, data have already demonstrated the weakness of the 'fortress conservation' position. We have already seen (Chapter 3) that isolated reserves act like 'land-bridge' islands, where formerly large areas within an environment suddenly become much smaller and more isolated. As a consequence, the movement of individual organisms is strongly reduced, and the populations living on the remaining habitat become much smaller. Since wild populations fluctuate a lot from year to year, small populations are more prone to extinction, and as there is little or no immigration, then these populations cannot be 'rescued' from extinction by the immigration of new individuals from elsewhere.

You have also already looked at the work of William Newmark on the parks of North America in Chapter 3. He showed that only the very largest of these parks was large enough to avoid 'natural' extinctions as a consequence of the island-like properties of nature reserves in the 'sea' of agricultural and otherwise human-influenced habitats. One possible response to this fragmentation of natural habitat is to join up the fragments via corridors. Again, you have already seen in Chapter 3 some of the evidence that such corridors can be effective. There are now plenty of studies, at a range of different scales, showing that corridors connecting patches of natural habitat do indeed slow down the extinction rates of animal and plant populations, and act as conduits for movement.

Not all animals use the corridors we provide, of course. We need to think carefully about the size and habitat type of the corridors, and the way particular target animals use their environment. For instance, cougars in Florida use the urban environment in quite a different way from that envisaged by conservation biologists, essentially ignoring the provided corridors and preferring to slink along routes they themselves have chosen.

How then will we learn to live with nature? The necessity for doing this is nowhere more evident than in the case of sparsely populated wide-ranging species such as large carnivores. The Cheetah *Acinonyx jubatus* is just such a species (Figure 8.3).

The cheetah is one of the most wide-ranging of all carnivores, with home ranges greater than $3000 \, km^2$, and movements of some animals exceeding $1000 \, km$. However, its densities are always very low, seldom greater than two per $100 \, km^2$, and sometimes one hundred times lower. Historically cheetahs were widespread across Africa and south-western Asia, but now they occur in only 9% of their former range (Figure 8.4) in a highly fragmented way. The total number alive today is estimated at 7100 adults and adolescents in 33 populations, more than half of which live in a single block of land stretching across six countries in southern Africa. Only one other population has more than 1000 individuals: most have 200 or fewer, and six populations contain fewer than ten individuals. Of 18 populations that could be assessed for trends, 14 were in decline, three were stable, and only one population was increasing. Of

Figure 8.3 Cheetahs are beautiful animals and the fastest land species, yet they are rapidly becoming extinct—future generations may never see these magnificent cats in the wild.

Francois van Heerden/Shutterstock.com

Figure 8.4 The current (red) and former (orange) range of the Cheetah *Acinonyx jubatus*. Protected areas are marked in blue.

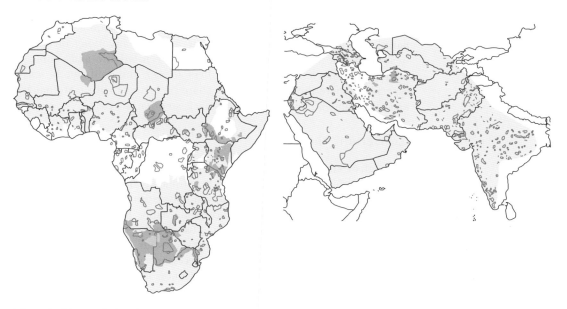

Adapted with permission from Durant, S. M. et al. The global decline of cheetah *Acinonyx jubatus* and what it means for conservation. *PNAS*. January 17, 2017 114 (3) 528-533. [https://doi.org/10.1073/pnas.1611122114]

all the Asian cheetahs, only the Iranian ones still exist, with fewer than 50 individuals in three populations—doomed to extinction, according to Iran's conservation authority.

To add to the problem, cheetahs also have very little genetic diversity. Cheetahs managed to survive the late Pleistocene extinctions caused by humans—but only in small numbers. This created a population bottleneck, and as a result the cheetahs alive today are all genetically very similar. This produces problems, because most cheetah crosses are effectively inbreeding. Many individuals have poor sperm quality, problems with their teeth and mouths, and kinked tails, and are susceptible to the same infectious diseases. All of these characteristics make it harder for the species to survive and do well, regardless of any conservation measures we may put in place.

So how are we going to save the cheetah? More than three-quarters of their current range lies outside Protected Areas (see Figure 8.4), where they come into conflict with local people by killing livestock. Their natural prey is lost to widespread overhunting and bushmeat hunting, their habitats are lost and fragmented, and they suffer from the illegal trade for pets or zoos. Individuals from areas with good populations do migrate to other areas, especially young adults, but most of these then die because there is nowhere for them to settle. The reasonable reproductive rates of some cheetah populations within Protected Areas are not enough to compensate for their declines in unprotected areas.

Here is a crystal-clear demonstration that Protected Areas are never going to be adequate for long-term survival. Growth rates inside Protected Areas have to be unrealistically high to compensate for the declines of the populations outside them (which contain two-thirds of the individuals). Conservation has to focus on increasing cheetah populations *outside* the parks. This in turn means a larger-scale approach involving incentives to local people to promote coexistence between wildlife and people in human-dominated landscapes. Needless to say, this will not be easy, especially where threatened species share their range with marginalized and vulnerable human communities, as is so often the case.

Conserving the oceans

Although we know less overall about the marine environment than we do about the land, marine conservation is in a better position than terrestrial efforts. This is because marine biologists have worked out systematically what is needed for marine Protected Areas to be successful. Marine parks across the world overall do a rather poor job at preserving fish density and fish species relative to nearby fished areas, but some are very successful. There are five elements that predict whether a marine park is successful or not. Success is defined as increasing the biomass or the species richness relative to the surrounding unprotected area (i.e. the predicted levels if the park were not there):

- The level of fishing allowed inside the Protected Areas needs to be low, at least in some areas; in particular, there have to be some no-take areas where fish are completely safe from exploitation by human fishers.

- Enforcement of the rules has to be efficiently done, with adequate policing as well as community support for the regulations; parks with good levels of enforcement show no obvious evidence of flouting of the regulations.

- Protection has to be in place long enough for the benefits to be realized: at least ten years are needed for these to appear.

- The Protected Areas have to be large enough; successful marine parks are more than 100 km² in area, whilst those less than 1 km² tend not to be successful.

- The fish habitat in the park has to be well isolated from surrounding fished areas so that movements of fish bred in the park are restricted. This allows fish populations of the marine park to be relatively independent of surrounding habitat. Parks where a reef runs continuously from inside to outside the boundary offer relatively poor protection, whilst those where the reef is contained within the park and isolated by a deep gulley or by a sand barrier do much better.

These five features are *additive*—no one element is sufficient on its own. If a park has only one or two of them, then its fish populations are no better off than the unprotected fished populations outside. However, the degree of protection rises rapidly when the number of elements increases from three to five (Figure 8.5), reaching very high values when all five are present: total fish biomass, for example, increases by almost 250%, and shark biomass by almost 2000%.

In the study there were only four marine Protected Areas that had all five elements, all small oceanic islands (belonging to Costa Rica, Colombia, New Zealand, and Australia). In addition, only five marine parks had four of the elements, and thus only 10% of the total sample of parks were regarded as effective. While many of these unsuccessful parks are small, and hence are never likely to achieve the criterion of being 'large', there is plenty of scope for them to achieve the other four.

The requirement for effective rule enforcement is interesting (Figure 8.6), because it is about the way the regulations work, which in turn is related to the extent to which local people 'buy in' to the idea of conservation. Places that are doing well are not just remote areas with low fishing pressure—they include areas with high human populations that make heavy use of the marine resources. These are places where there are strong local cultural institutions associated with the use of marine resources. They include rules of tenure that allow exclusion of fishers from outside the local village, and restrictions on resource use through an adaptive rotational harvest system based on what they catch.

Figure 8.5 The difference in fish biomass (upper) and the species richness of larger fish species (lower) in marine Protected Areas with different numbers of key elements (see text). The y-axis is the percentage change from the value predicted if the park did not exist.

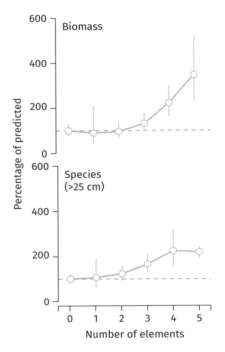

Adapted from Edgar G. J. et al. Global conservation outcomes depend on marine protected areas with five key features. *Nature* 506: 216-220. Copyright © 2014, Springer Nature: https://doi.org/10.1038/nature13022

Crucially, a high level of local engagement and control of the management process is necessary. Areas near large and accessible fish markets are not doing well in conservation, and neither are those that have imported modern fishing technologies, such as fish freezers and destructive netting, that can lead to higher levels of exploitation. Markets create incentives for over-exploitation not only by influencing price and price variation for the products of reefs, but also by influencing human behaviour, especially the level of willingness to cooperate in the collective management of natural resources.

Thus sustainability will happen where local people depend on local resources for their own nutrition, have control of the management of those resources, and have traditional mechanisms to prevent over-exploitation, such as rules of land tenure and of rotational use to allow the resource to recover (see Figure 8.6). Applying these principles to countries where high levels of technology are the norm raises all sorts of issues.

So are these ideas only relevant to less-developed countries? Of course not! With the scaling up of our understanding of the drivers of the species richness and the structure of ecological communities comes the knowledge that many mechanisms occur at landscape scales, or even nationwide, continent-wide,

Figure 8.6 Differences in key social indicators between areas with good, average, or poor levels of fish biomass/species richness. The y-axis is the proportion of sites that show each feature. Sustainability happens where there are local traditional mechanisms of fishing tenure rights and rotation of fishing to allow recovery, where there is strong local control of management, and where local people depend on fish for their livelihoods. The occurrence of funded projects from outside has no discernible effect.

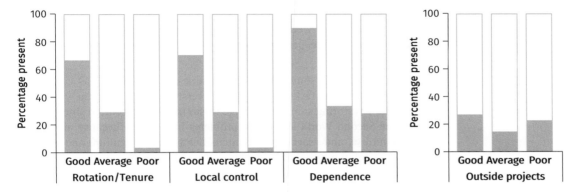

Adapted from Cinner, J. E. et al. Bright spots among the world's coral reefs. *Nature* 535: 416-19. June 15 2016. Copyright © 2016, Springer Nature: https://doi.org/10.1038/nature18607

or planet-wide scales. A good example is the weather. It is simply impossible to understand and predict the weather without a detailed planet-wide model of the atmosphere. Even then it can only accurately be predicted a few days ahead because of the complexity and scale of the processes involved, coupled with the operation of chance, errors of measurement, and ignorance of some important elements. In the same way, we will not achieve the sustainable conservation of biodiversity by setting aside some nature reserves, and imagining that the job is done.

UK conservation

We have already (Chapter 3) mentioned Professor Sir John Lawton's survey of the state of nature conservation in the UK, *Making space for nature*. It is a sobering read for a nation that prides itself on its attitude, knowledge, and care for nature. It starts with a fascinating analogy:

"There are twenty-seven ancient cathedrals in England. Imagine the outrage that would have ensued in this country if over the last 100 years, twelve had been partly demolished, nine substantially demolished, and three completely obliterated; only three would remain in good condition. Yet this is precisely what has happened to many of England's finest wildlife sites.

Between 1912 and 1915, Charles Rothschild, a banker and naturalist, conducted a survey of sites of conservation importance in England, with a view to setting up a nationwide network of nature reserves. The list . . . had 182 sites in England . . . The basis on which sites were selected is not perfect . . . , but their subsequent fate is illuminating. In the last detailed analysis [1997], led by Rothschild's daughter, . . . the score was:

- Little or no loss: 19 (10%)
- Less than 50% loss of the habitats for which it was originally listed: 84 (46%)
- More than 50% loss: 58 (32%)
- Total loss: 21 (12%).

These are (or were) the cathedrals of nature conservation, yet we see no great outrage, perhaps because so few people realise what we have so easily lost. "

Lawton's team found Britain's network of Protected Areas completely inadequate to the task of keeping biodiversity at its level in the year 2000, let alone restoring some of the huge losses that have occurred since the end of the Second World War. Most problems have arisen because of changes in land use, and especially the intensification of agriculture and fertilizer use. Under the guise of efficiency in agriculture, Britain's habitats have suffered and the specialist plants and animals that depend on them have declined or disappeared, leaving just the 'weedy' common generalists. Over various timescales the UK has lost nearly all of its fens (98%), species-rich grasslands (97%), grazing marsh (81%) and lowland heaths (80%). Britain was once largely covered by ancient woodlands, yet nearly all had been lost by 1600. Of what remained, a further 7% has been lost since 1930, with much of the rest converted into conifer plantations.

There is some good news, in that the existing set of wildlife sites does contain a representative set of UK habitats and covers a good proportion of its remaining biodiversity. However, the sizes of Protected Areas in the UK are much too small—most of them are less than 100 hectares—and many are not funded or managed adequately. Small reserves are more strongly affected by the quality of the surrounding land, and cannot contain self-sustaining populations unless managed extremely carefully. Furthermore, most of the natural connections between conservation sites, such as hedgerows, ponds, and rivers, have been lost.

If implemented, the recommendations of the Lawton report would be radical because they require much more land to be devoted to wildlife than at present, and coherent networks with wildlife-friendly corridors to be established. The three objectives are:

- to restore species and habitats to levels enhanced over those existing in 2000;
- to restore and make sustainable the ecological processes that underpin ecosystem services of our natural environment, such as clean water, climate regulation, and crop pollination;
- to provide accessible nature rich in wildlife for people to experience and enjoy.

Effectively the report calls for a terrestrial version of the elements of the marine parks (Figure 8.7), where Protected Areas are large enough, well enough managed, with local 'buy-in' and control, connected and buffered against other kinds of activities. They should have core areas of high conservation value where the highest concentrations of rare and/or specialized species can thrive then disperse to other parts of the network. New areas of natural habitat will need to be restored or created that will eventually become new core areas, and these should be situated so as to complement, connect, or enhance the existing core areas.

Different parts of this strategy will benefit different kinds of organisms, mostly in relation to their mobility. The evidence shows that plants will benefit most from localized management that includes reductions in the intensity of agriculture. In contrast, mobile vertebrates such as birds don't really respond to such localized efforts, but instead benefit most from a landscape-scale approach that maximizes landscape complexity. Invertebrates such as bees respond to management efforts at both scales.

Thus people in the UK, one of the world's richest countries, have one of the hardest of tasks to create sustainable conserved ecosystems and ecosystem processes. Just as with most major infrastructure projects, the costs of conservation projects in the UK will be three times higher than the same work would be in countries such as Germany or France. This is simply because our population density is three times greater, hence we have so much more to remove in order to re-engineer the landscape. In spite of this, Lawton's vision of an ecologically sustainable UK is achievable, if the general public can be persuaded of the need, and the political will is there.

Figure 8.7 Planning improvements to Britain's conservation network.

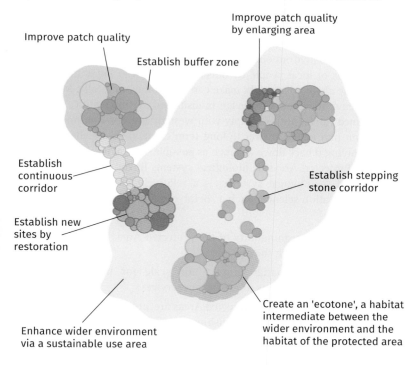

Adapted from Lawton, J. H. et al., 2010, *Making Space for Nature: a review of England's wildlife sites and ecological network.* Report to Defra. Contains public sector information licensed under the Open Government Licence v3.0.

Human wellbeing and delivering the goal

How are we to deliver the goal of transforming landscapes to serve both human and conservation interests? At the moment, the only thing proven to increase human health and wellbeing in all contexts is conventional economic development, which nearly always pays no heed at all to conservation, sustainable or otherwise. The hard question is whether conservation can deliver benefits that match. Is this possible?

If you take the 'social conservationist' viewpoint, then you believe it possible to link conservation and development in order to alleviate the poverty and disadvantage that is normal for the people who live in and around areas of conservation importance. Social conservationists seek ways to persuade, consult, and engage local people with the problem of how to conserve wildlife without damaging their livelihoods (or replacing them if change is unavoidable). At a very different level, this is also a dominant theme of the Convention on Biological Diversity—to persuade decision-makers that conservation needs to be part of mainstream thinking, incorporated into all aspects of human life, including all policy decision-making.

Local people will only agree to cooperate in conservation initiatives if they trust those putting them forward, and if it makes sense to them. Projects need to benefit individuals, their families, or their communities either directly or indirectly—there are obviously many different possible motivations. As we have seen, often in the past local societies worked out effective ways of managing the resources that matter to them. However, more and more power has become centralized in the hands of modernizing governments and international organizations, whose thinking is grounded in western scientific models to the exclusion of alternative outlooks. Control of resource management is wrenched from the hands of local people by outside 'experts' who know their subject but know little about what might make conservation work locally. Meanwhile traditional environmental knowledge is dismissed as unscientific, and local people blamed for failure to comply with western conservation norms. Thus an important part of the answer to long-term conservation is about shifting power, devolving it back down as much as possible to local levels.

Treating Protected Areas as biological systems linked with local people, i.e. with local sociological systems, is a relatively recent but increasing view in conservation science. Such a social-ecological approach considers three aspects:

- the long-term sustainability of the Protected Areas themselves in the landscapes in which they occur;
- the appropriate spatial context and scale for them to function effectively as producers of ecosystem services;
- the redefinition of what Protected Areas are, and how they both define and are defined by the relationships between people and nature.

This kind of conservation is quite different from what has gone on in the past. It recasts biodiversity conservation as the search to understand the interactions between people and nature, to co-operate with and influence them, and to create the appropriate enabling conditions for them to flourish whilst achieving acceptable ethical, equitable, and political solutions for all the diverse stakeholders.

Some government-led top-down initiatives seem to work in engineering sustainable solutions—one example is payment for ecosystem services, i.e. the government through the conservation authorities pays local people to act in an environmentally friendly way. In some countries this appears to work well. For example, villagers in the 2000-km² Wolong Nature Reserve in southwestern China that protects the largest population of Giant Pandas (Figure 8.8) are compensated for not doing activities that damage panda habitat, such as collecting firewood. Local people must have confidence that such compensation will happen reliably—a relationship of trust. This is part of an integrated approach involving scientific research, management, and a thriving captive population of pandas. Wild panda numbers have increased by 17%, and their conservation status has been downgraded from Endangered to Vulnerable. There are now 67 panda reserves protecting two-thirds of wild pandas and 1.4 million hectares of panda habitat. Of course, these also protect a myriad other animal and plant species—a great example of an 'umbrella' species whose targeted protection conserves biodiversity in general.

Figure 8.8 A combination of conservation and captive breeding in the Wolong Nature Reserve in China, strongly supported by the local population, has led to the beginnings of a revival in numbers of this iconic species.

Danita Delimont/Alamy Stock Photo

The way forward

The last two decades of conservation research suggest that the needs of humans and nature go hand in hand, and achieving sustainability requires a consideration of both sides.

At a global scale, international action is needed by countries acting together with the common aim of preserving ecosystem services upon which all human life depends. The two main conventions signed by most countries are the 1973 Convention on International Trade in Endangered Species (CITES) and the 1992 Convention on Biological Diversity (CBD). These have been very effective in channelling the efforts of individual countries in a common direction. The CBD is now into its second round of Biodiversity Targets, goals to be achieved by 2020 (Table 8.2).

An effective agreement on tackling climate change is also vital. The science of global warming is well understood, and is well communicated to governments and the public by the Intergovernmental Panel on Climate Change (the IPCC). The Intergovernmental science-policy Platform on Biodiversity and Ecosystem Services (IPBES), established in 2012, aims to perform the same invaluable role for biodiversity and conservation.

At a country scale, many countries have commissioned plans for improving the conservation of biodiversity and ecosystem services within their borders, often as a direct response to the targets of the CBD to which they have signed up. We have already discussed the UK's version, the Lawton Report *Making space for Nature*. It's all very well having plans, but the challenge is ensuring that governments enact them.

Breaking out of the habit of land appropriation and short-term exploitation towards sustainable development of human communities will involve a different way of working, one that promotes the trust of local people and devolves power down to the local level. This is especially true for indigenous people, because it must explicitly value their traditional knowledge and customary laws. Connections between conservation donors (individuals, organizations, and governments) and local people must shift power away from the top-down approaches towards bottom-up control by local communities, in just the way that the management of marine

Table 8.2 The Biodiversity Targets of the Convention on Biological Diversity.

Strategic Goal A:	Address the underlying causes of biodiversity loss by mainstreaming biodiversity across government and society (targets 1–4)
Strategic Goal B:	Reduce the direct pressures on biodiversity and promote sustainable use (targets 5–10)
Strategic Goal C:	To improve the status of biodiversity by safeguarding ecosystems, species, and genetic diversity (targets 11–13)
Strategic Goal D:	Enhance the benefits to all from biodiversity and ecosystem services (targets 14–16)
Strategic Goal E:	Enhance implementation through participatory planning, knowledge management, and capacity building (targets 17–20)

(see www.cbd.int/sp/targets)

biodiversity has shown is required. Relationships of trust and transfers of power are always difficult to achieve, but one possible transferrable model may be found in community philanthropy organizations, a localized approach increasingly recognized as effective in empowering local people to achieve wellbeing.

A community philanthropy organization is based in and run by and for people in a local community. It encourages local people to contribute resources of money, time, and materials, however limited, to promote community wellbeing, a 'good society' and social justice, and to defend against intrusions by the state on individual rights and freedoms. By remaining local and small, such organizations avoid the issues associated with being large and wealthy, which almost always entail joining the 'establishment' and becoming part of the problem rather than the solution. The priorities of organizations with large sums of money—whether governmental or private—are usually conservative, and very far from those of the recipients of their aid. By contrast, small community organizations are close to the ground and can survive only by addressing the priorities of, and securing the trust and the buy-in of, local people.

How should we define the 'wellbeing' that these local organizations aspire to foster? According to a growing literature it's definitely not equivalent to economic growth—the objective of development is not to reproduce London in all the towns and cities of the world. It lies in the capacity of individuals to make free choices to improve their lives *according to their own norms and values*. Even the capacity to aspire to make improvements will make a difference to how people feel about the quality of their lives.

Wellbeing is not the same as 'happiness': people will often say that they are happy even though their personal circumstances would be considered very grim by someone else. Stated happiness is often less a reflection of whether or not a person has realized their hopes than of having learnt to live within the limits imposed by their circumstances.

We can measure wellbeing using the concept of *agency*, i.e. the ability, opportunity, and freedom to engage in community life. There are five elements:

- the ability to access physical, material, and intellectual resources;
- feelings of competence and of self-worth;
- the ability to participate in whatever seems important;
- the ability to sustain social connections;
- the ability to achieve physical and psychological health.

The ability to exercise agency is affected by power relationships, identity, and influence, and can be facilitated or blocked by the social and political environment. All of the elements are profoundly centred on *place*, i.e. the community in which people live out their lives. One of the most fundamental elements of human wellbeing lies in the quality of social relationships, which simply cannot be promoted by outsiders without a deep understanding of the lives and circumstances of people in a given place. Once such agency is enabled, then people can more effectively shape their own destinies and environment with less need of external inputs.

Thus one important element to achieve sustainability in many areas of conservation interest across the world is long-term partnerships with locally focused, locally led, highly networked community organizations. Such organizations are usually considered by donors to be too small to be significant players, but they are best-placed to effect sustainable change. They are rooted in *place*, with deep knowledge of the systems, issues, and conditions that affect residents, and an ability to act in response to them. They support appropriate local development initiatives, enabling them to be owned, repeated, and spread. They have the trust of local people, developing networks of relationships and resources that span the whole community, and their priorities are shaped by the local community rather than by outsiders. It is with such grassroots networks of community knowledge that conservation initiatives should think of aligning themselves, if they are to become locally rooted and adopted by the community as relevant and appropriate to its needs. In this way a learning community can be created, in which conservationists learn from local people how their environment works (or doesn't), while sharing new scientific perspectives that may enhance its health. Whilst partnering with such organizations may sometimes involve risk, their approach offers hope for supporting an effective conservation model that puts the needs and expertise of local people at its heart.

Case study 8.1
Harmonizing food security and biodiversity conservation: soybean production in the Brazilian Cerrado

Our final case study comes from work by Yvette Perfecto and her colleagues in Brazil, and represents an attempt to bring together the goals of agricultural production, biodiversity conservation, and the wellbeing of local people. Agriculture need not entail ecosystem degradation and reductions in biodiversity, and both can be achieved while simultaneously providing local social justice and empowerment. These intersecting processes also interact with scale effects because they are embedded in the governmental regulatory and policy environment, mediated by institutions.

The Cerrado is a vast savanna region of Brazil, encompassing a number of states. The work was done in the state of Mato Grosso, a sparsely populated area where more than 50% of the GDP comes from agriculture. Despite rapid economic growth, 15% of people remain below the poverty line and almost 20% experience food insecurity. The transition to agriculture has contributed to widespread deforestation. There are some large-scale farms with highly industrialized cropping systems and cattle ranches, but more than 80% of farms are owned by smallholder families. Crudely speaking, there are two types of landscape: large-scale intensive producers of soybean and beef for export that require external inputs of seed varieties, fertilizers, and pesticides; and diverse family farms focused on domestic markets and relying on local inputs of legumes and animal manures.

Encouraged by national and state support, Mato Grosso contributes about a quarter of Brazil's soybean exports from farms averaging 3000 ha. At both landscape and local scales, the wealth from these farms is concentrated among small numbers of producers and agri-businesses, contributing little to local economic development, local equality, or environmental sustainability. The construction of soybean farms causes deforestation and biodiversity loss, and when fully operational they are characterized by low biodiversity, high dependence on fossil-fuel-based inputs, and pesticide contamination of the water supply. These wealthy people and companies maintain favourable institutions and disrupt others that might actually generate better collective outcomes for biodiversity and food security. This is the problem of 'elite capture' of the reins of power and governance, a common problem all over the world.

In contrast, the 85 000 family farms in Mato Grosso average just 50–70 ha, producing mixed grains, vegetables, fruit, and livestock for local consumption. Most of the work is done by family labour, and there are 4–5 jobs per 100 ha compared to 0.3 in the large soybean farms. Family farms were strongly influenced by social movements for agrarian reform in Brazil that sought more equal access to rights and resources to sustain rural livelihoods, and thereby increase food security. These movements were successful in encouraging farmer-led market cooperatives and general support for diverse, locally based, ecologically sound markets. The farms are integrated with the surrounding landscape such that both wild and agricultural biodiversity are higher than in large soybean farms, although not enough research has been carried out to be sure just how much higher they are.

Small-scale production for family subsistence and for sale in local markets ensures domestic food security, as well as a diverse and high-quality diet. Such markets are less affected by global events and speculators. A diverse mix of crops makes households less vulnerable to market volatility and ecological risks such as weather or disease.

The contrasting landscapes are summarized in Figure A. Although this study does not mention the explicit involvement of community organizations, it is easy to see where such inputs might help. For example, a local community philanthropic organization might set up to support and be supported by members of the farmers' cooperatives.

Although just starting, this kind of analysis is part of an integrated view of ecosystem services and the wellbeing of local people. It is critically important if we are to develop a sustainable environment at all levels, from local to global. There is not necessarily any conflict between biodiversity conservation, agricultural production, and human wellbeing. The trick is to find the win-win-win solution.

CS8.1 Figure A Scale and social-ecological systems in the Mato Grosso of Brazil.

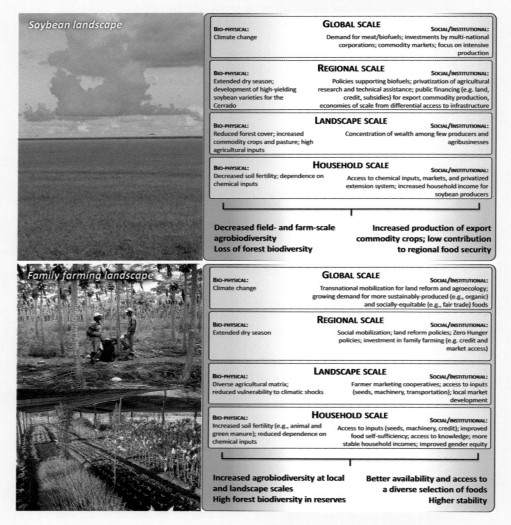

Reproduced from Regional Environmental Change. Hannah Wittman, Michael Jahi Chappell, David James Abson et al. *A social–ecological perspective on harmonizing food security and biodiversity conservation*. June 2017, Volume 17, Issue 5, pp 1291–1301. Springer Berlin Heidelberg.

 **Pause for thought**

How might you apply such thinking to a developed country such as the UK, where so few people work as farmers?

Chapter summary

- There is an important debate on the best way to conserve ecosystems. Nature protectionists insist that people-free Protected Areas are the only proven method, whilst social conservationists insist that the sustainable use of wildlife with social justice for local people is the only long-term solution.

- Most wildlife is not contained within Protected Areas, so people need to learn to live with nature, especially because of the ecosystem services that it provides.

- Evidence shows that five elements need to be present for marine Protected Areas to be effective, including the engagement of local people. Similar design rules are probably relevant to the effectiveness of terrestrial Protected Areas as well.

- Global international conservation has the right priorities, but enacting them is hard; for example, the plan for UK conservation calls for a wildlife-friendly landscape, with a large expansion in reserve area and connectedness, together with more effective management.

- Human wellbeing is embedded in the local community, where *agency* is critical—the ability to make choices based on one's own prioirities. There is not necessarily any conflict between biodiversity conservation, agricultural production, and human wellbeing. The trick is to find the win-win-win solution.

Further reading

Miller TR, Minteer BA & Malan LC (2011)

The new conservation debate: the view from practical ethics. *Biological Conservation* 144: 948–957. This paper outlines the great debate about how to do conservation.

Wittman H et al (2017)

A social-ecological perspective on harmonizing food security and biodiversity conservation. *Regional Environmental Change* 17: 1291–1301. A paper that expands on the case study of this chapter.

These two websites contain a lot of information about endangered species and about the international effort to conserve biodiversity:
www.iucnredlist.org
www.cbd.int

Discussion questions

8.1 Where do you stand in the great conservation debate? Outline both sides in the argument and explain your views.

8.2 Find examples of threatened ecosystems in two different areas of the world, and for both suggest ways in which we can engineer the action that effective conservation requires.

Glossary

abiotic the non-living elements of the environment, e.g. temperature, soil moisture, etc.

adaptation this can refer either to a *state* (the end-point of the process of natural selection) or a *process* (the way natural selection causes organisms to be better fitted to their environment)

adaptive radiation the process by which a single species colonizing a new environment speciates into many species, each typically occupying very different niches

alien species an alternative name for *invasive* species introduced by humans in the colonization process (and therefore 'artificial')

alkaloid a class of chemicals produced by plants to defend themselves against herbivores

allele one form of a gene on a chromosome

Anthropocene a recent name for the latest period of the Earth's history that contains human beings

apocrine gland glands whose secretion is released into the duct by the dissolution of half of the cell; they are limited to the armpits and anal region of humans, but are the normal sweat glands of many ungulates

arithmetically typically something that increases by addition, e.g. 2, 4, 6, 8,...

biodiversity a general term for the number and range of types of organism in the living world

biomass the total mass of living things in a specified area

biotic the living elements of the environment

biotic filtering this refers to the way in which organisms are prevented from colonizing a habitat ('filtered') because of the kind of organism they are, e.g. islands cannot support large mammals

bottom-up control communities where populations are controlled by the availability of their resources (typically food), i.e. from below in the trophic levels

brackish water water containing salt concentrations less than full-strength seawater, but more than freshwater

carrying capacity the limit to the population density that the resources of the environment can support

clade one branch (+ twigs) of a phylogeny (an evolutionary tree)

coevolutionary arms race the process of mutual co-adaptation between a prey and a predator, or between a herbivore and a plant

community the set of species in a habitat linked by feeding relationships in a food web

coral bleaching the process that occurs at high seawater temperatures whereby coral expels its symbiotic algae, the zooxanthellae

cyanogenic glycoside a class of chemicals produced by many plants to defend themselves against herbivores

demographic processes the processes of birth, death, emigration, and immigration that alter the numbers of individuals in a population

density-dependence the name for processes regulating population numbers that are related to population density

dispersal limitation the process whereby species are absent from a habitat because they are unable to reach it via movement

dynamic equilibrium A system which remains the same as the result of constant compensatory changes

eccrine gland a gland that secretes a product by putting it in a vesicle that merges with the cell membrane, expelling the secretion into a duct; most human sweat glands are eccrine glands

ecological footprint the impact of human beings on the Earth's environments

ecologist a scientist studying how populations and communities 'work'

ecosystem a rather vague collective term for sets of similar habitats, e.g. the desert ecosystem

ecosystem services the set of benefits provided to human beings by the natural world

ectothermic organisms that rely only or mainly on external heat from the environment for their activities

endemic an organism that only occurs in one region of the world is endemic to that region

endothermic organisms that generate their own metabolic heat for their activities

Enlightenment the period of history in the seventeenth and eighteenth centuries when science and reason took over from religion as the main ways of understanding the world

environmental filtering the way organisms are prevented from colonizing a habitat ('filtered') because the environment they need is not there, e.g. an island without trees cannot support tree-living monkeys

exclosure a cage designed to exclude certain organisms or processes (e.g. excludes grazing animals)

extinct the death of the last individual of a species makes that species extinct

extinction debt when a habitat or island is made smaller, inevitably some species will die out but this may take time; the extinction debt is the number of species doomed in the long run to die out from an area

fitness a key concept in ecology and evolution, the fitness of an individual female is the number of offspring she has over her lifetime that survive to reproduce themselves

fragmentation this occurs when a habitat becomes fragmented into several smaller more isolated pieces

functional explanation this is one of Tinbergen's 'Four Why's', where the answer to why a trait exists is concerned with its function

functional group a set of species that all make similar demands of the habitat, e.g. all the herbivores that get their food from eating plants

fundamental niche where (in terms of abiotic conditions) a population would live if only the abiotic environment determined its fitness

genetic drift the way chance can change the genetic makeup of a population, especially when it is small

geometrically something that increases by multiplication, e.g. doubling each step: 2, 4, 8, 16, 32,...

glucosinolate a class of chemicals, typically produced by brassica plants to defend themselves against herbivores

habitat a rather vague term in ecology for a place with similar abiotic or biotic conditions where particular organisms live

holism an approach to science that acknowledges that groups can have properties that are not explicable from the properties of the individuals that make up the group

human-induced climate change the way climate is changing extremely rapidly since the 1970s is clearly caused by human industrial activities

intransitive the rock-scissors-paper game in ecology; i.e. if A beats B, and B beats C, then you might expect A to beat C ('transitive'), but if this does not happen (C beats A) then the relations are called 'intransitive'

invasibility the extent to which a community is capable of being invaded by other species

isotherm a line on a map showing where the same average temperature has been recorded

limiting similarity the limit to how similar two species can be before they compete too strongly to coexist in the same community

manipulation experiment the gold standard in scientific proof, where the scientist experimentally manipulates one factor and observes the effect of this on another

mass effects where species appear to coexist in a habitat simply because migrants continually arrive there

maternal effect non-genetic inheritance from mother to offspring contained in the cytoplasm of the egg; typically such effects can be seen in the offspring, and sometimes the second generation, but are not inherited further

mechanistic explanation this is one of Tinbergen's 'Four Why's', where the answer to why a trait exists is concerned with the components from which it is made

metapopulation all populations are patchy in the real world, so a metapopulation is a set of patchy populations connected by movement; in the strict sense, it is a set of extinction-prone patch populations that are maintained by immigration

mutation any process that causes changes in the DNA of a gene; most mutations are caused by uncorrected mistakes in DNA replication

mutual dependence a relationship between species where each depends on the other to remain in a particular community

mutualism an evolved relationship between species where each benefits the other

natural enemies a collective term for all the predators, parasites, and parasitoids of a particular species

nature protectionists people who believe that only nature reserves without people can provide guaranteed success in conservation

net primary productivity the total net carbon fixed by all photosynthesizing organisms

network a general term for a set of things connected by relationships

niche the set of resources that an organism requires in order to sustain its populations

ontogenetic explanation this is one of Tinbergen's 'Four Why's', where the answer to why a trait exists is concerned with how it develops from the egg to the adult organism

paradigm the underlying set of assumptions that underpin how we interpret the world; in biology, our main paradigm is the theory of evolution by natural selection

parasitoid an insect that lays its egg in another insect, and its larva eats the tissues of the host so completely that it dies

patch a discrete set of resources separate from others

phenolic a class of chemicals produced by many plants, including many trees, to defend themselves against herbivores

phenology the pattern of the appearance of an organism during the year

phylogenetic explanation this is one of Tinbergen's 'Four Why's', where the answer to why a trait exists is concerned with the evolutionary pathway from its first ancestral appearance

polyp the individual organism in a coral (which is a colony); it has a stem and a circle of tentacles

population dynamics the changes in the numbers of individuals in a population through time

potential evapotranspiration evaporation is the combined water loss from the soil from evaporation and transpiration by plants; potential evapotranspiration is the total water demand if there is enough water in the soil to satisfy it

Principle of Competitive Exclusion the idea that some species compete too strongly to be able to coexist in a community

Protected Areas the general term for officially designated (hence the capitals) areas for conserving wildlife

realized niche the part of the fundamental niche left to an organism once competitive exclusion has been taken into account

recruitment this describes the set of juveniles that survive to become reproducing adults (new 'recruits') in the population

Red Queen the idea that all components of the biotic environment are coevolving together

reductionism an approach to science that suggests that all properties of an object are explicable from the properties of the bits that make it up

regulation in the context of populations, regulation means that numbers are prevented from exceeding particular lower and higher limits by mechanisms which an ecologist sets out to discover

resource anything required by an organism and finite in availability

Romanticism an eighteenth- and nineteenth-century movement that exalted subjectivity, emotion, and Nature, and downplayed objectivity, science, and reason

ruminant a group of mammals that possess a rumen, i.e. a chamber in the stomach where the food is fermented before being chewed a second time before digestion

saturated community a community that resists invasion by other species

social conservationists people who believe that sustainable conservation will only happen by involving local people and their sustainable development via poverty alleviation

social-ecological approach the new science of combining ecology and social science in conservation

symbiotic an evolved relationship between organisms where each cannot survive without the other

taxon cycle a long-term process of range expansion, followed by genetic differentiation then speciation and extinction, said to explain many distribution patterns today

terpene a class of chemicals produced by many plants, including many trees, to defend themselves against herbivores

top-down control communities where populations are controlled by their natural enemies (typically predators or parasitoids), i.e. from above in the trophic levels

trait a particular feature of an organism

trophic cascade the cascading effects of one species through a food web, usually from a top predator down though the web, but sometimes reversed from a plant upwards

Tullgren funnel a method of sampling soil and moss organisms involving a light bulb over a grill on top of a beaker containing preservative; the light dries out the sample, and in trying to escape the organisms fall through the beaker into the preservative; it is very effective

understorey the herbs and shrubs that grow underneath a tree canopy

ungulate the group of mammals that have hooves, possibly not a natural group (i.e. a group containing unrelated clades)

zooxanthellae symbiotic algae within the bodies of corals that fix carbon by photosynthesis and pass it to the coral

INDEX